The Reference Shelf®

U.S. National Debate Topic 2011–2012

American Space Exploration and Development

Edited by
Christopher Mari

The Reference Shelf
Volume 83 • Number 3
The H.W. Wilson Company
New York • Dublin
2011

The Reference Shelf

The books in this series contain reprints of articles, excerpts from books, addresses on current issues, and studies of social trends in the United States and other countries. There are six separately bound numbers in each volume, all of which are usually published in the same calendar year. Numbers one through five are each devoted to a single subject, providing background information and discussion from various points of view and concluding with a subject index and comprehensive bibliography that lists books, pamphlets, and abstracts of additional articles on the subject. The final number of each volume is a collection of recent speeches, and it contains a cumulative speaker index. Books in the series may be purchased individually or on subscription.

Library of Congress has cataloged this serial title as follows:

U.S. national debate topic 2011-2012 : American space exploration and development / edited by Christopher Mari.
p. cm. -- (The reference shelf ; v. 83, no. 3)
Includes bibliographical references and index.
ISBN 978-0-8242-1109-7 (alk. paper)
1. Astronautics--United States. 2. Outer space--Exploration. 3. United States. National Aeronautics and Space Administration. I. Mari, Christopher. II. Title: American space exploration and development. III. Title: United States national debate topic 2011-2012. IV. Title: US national debate topic 2011-2012.
TL789.8.U5U514 2011
629.40973--dc23
2011014617

Cover: This artist's concept depicts NASA's Phoenix Mars Lander a moment before its 2008 touchdown on the arctic plains of Mars. Pulsed rocket engines control the spacecraft's speed during the final seconds of descent. Image Credit: NASA/JPL-Calech/University of Arizona.

Visit H.W. Wilson's Web site: www.hwwilson.com

Printed in the United States of America

Contents

IV A Bold Mission or Sheer Lunacy? Should We Return to the Moon?

V Greenlighting the Red Planet: Should We Go Straight to Mars?

Appendix

Bibliography

Preface

The U.S. space program and the agency that runs it, the National Aeronautics and Space Administration (NASA), are at a crossroads. While NASA's robotic exploration of the solar system and beyond has succeeded brilliantly for decades, providing researchers with tremendous insights into the nature of our universe, the U.S. manned space program has been locked in low Earth orbit (LEO) since the early 1970s. When *Apollo 17* lifted off from the moon in December 1972, America effectively abandoned the human exploration of other worlds, calling it quits after just six manned lunar landings.

This begs the question: What happened? Given mankind's boundless curiosity and tendency to push the technological envelope, it seems odd that we were once capable of sending astronauts to the moon—and potentially other planets, such as Mars—but now no longer can. It's especially perplexing when one considers that NASA's "brand"—its very identity—centers on manned spaceflight. Alan Shepard's first suborbital flight in 1961, John Glenn's first orbit in 1962, Neil Armstrong's first walk on the moon in 1969: These missions defined NASA for a generation of Americans. Despite the enormous scientific value of NASA's unmanned probes, it's the image of human beings blasting off into the cosmos that thrills young and old alike. There's nothing like a human footprint on a foreign world to make one feel humbled, inspired, and awestruck, all at the same time.

Yet for the last 30 years, NASA's efforts have centered on the space shuttle program, officially the Space Transportation System (STS). The space shuttle can ferry astronauts to and from Earth orbit, stopping at the International Space Station (ISS) and Hubble Space Telescope, but it can't travel to other planets. The program has proven the feasibility of a reusable spacecraft, but it's also suffered two catastrophic losses: the in-flight explosion of the space shuttle *Challenger* in January 1986 and the breakup of the space shuttle *Columbia* upon reentry in February 2003. Both accidents raised questions about the safety of reusable launch vehicles, and in 2004, in the wake of the *Columbia* disaster, President George W. Bush advocated canceling the space shuttle program and bringing back non-reusable spacecraft, like the ones employed in the original Apollo lunar missions.

Bush's plan, the Vision for Space Exploration, called for the completion of the International Space Station by 2010, the termination of the space shuttle program the same year, and the development of a new "Crew Exploration Vehicle"—

later dubbed Orion—by 2008. This new spacecraft was to have conducted its first manned space mission by 2014 and landed human beings on the moon by 2020, this time with the express purpose of establishing a permanent human presence there. The skills, experience, and technical expertise developed on these new lunar missions would then have been used to conduct a sustained exploration of Mars, via a program known as Project Constellation.

Unfortunately, Constellation's future turned out to be not so bright. In recent years, in the midst of the worst economic recession since the Great Depression, a host of circumstances—insufficient funding, cost overruns, technical setbacks, and an unwillingness on the part of Congress and President Barack Obama to allocate more money—ultimately conspired to kill the program. Obama instead asked Congress to fund a scaled-down plan that would rely on the private sector to create the next generation of American spacecraft—vehicles capable of going into orbit or visiting nearby asteroids. In the meantime, Obama planned to send U.S. astronauts to the ISS by paying for seats aboard Russia's Soyuz spacecraft.

As of this writing, the space shuttle program is slated to end in 2011, and a replacement won't be ready until 2015 or 2016. Further, NASA appears to have no clear mandate to explore the moon or Mars in the near future, although President Obama has said, "By the mid-2030s, I believe we can send humans to orbit Mars and return them safely to Earth. And a landing on Mars will follow. And I expect to be around to see it." Without specific goals and the funds to reach them, such vague assurances have left many proponents of space exploration—including NASA employees and members of Congress on both sides of the aisle—frustrated and bewildered.

By asking private companies to design new U.S. crew vehicles, NASA would move into uncharted territory. Such a course would raise many questions: Would these new vehicles be as reliable as the ones NASA has commissioned and overseen throughout its history? Would private companies adhere to the same rigorous safety oversights NASA employs?

Clearly, NASA is no longer the agency it was in the 1960s, when it enjoyed a firm directive—President John F. Kennedy's challenge to reach the moon by decade's end—and boasted a laundry list of concrete accomplishments. The image of Neil Armstrong taking mankind's first steps on the moon in July 1969 is no longer a symbol of progress; it's a museum piece. The NASA of the future will not be the NASA of old or even the NASA of the present, with its highly successful unmanned exploration program. But what will it be?

The U.S. National Debate Topic 2011–2012 declares: "Resolved: The United States federal government should substantially increase its exploration and/or development of space beyond the Earth's mesosphere." Few topics could be timelier. U.S. space policy is entering a new phase. Will the U.S. government rededicate itself to space exploration, or will it let other nations and private aerospace companies blaze new paths beyond our blue world? Does a robust space program, whether manned or unmanned, represent a nation that is progressive and optimistic about its future? Conversely, does a diminished space program suggest a country less willing to take risks? Do the technical and economic benefits of space exploration outweigh its inherent risks? Does maintaining the space program waste money that could be used to solve more pressing problems here on Earth?

The articles collected in this volume of The Reference Shelf are intended to give debaters and general readers alike an overview of some of the critical issues facing U.S. space exploration. Entries in the first section, "The U.S. Space Program: A Laudable Past, a Muddled Present," provide a history of NASA from the Cold War to its current moment of crisis. Pieces in the next chapter, "At a Crossroads: Should NASA Be Overhauled?" debate the pros and cons of President Obama's new plan. Selections in "Outer Space Goes Out to Bid: Should Private Companies Lead U.S. Spaceflight?" profile several of the private companies building launch vehicles for NASA. Although the Obama administration has no plans to return to the moon, articles in "A Bold Mission or Sheer Lunacy? Should We Return to the Moon?" consider whether we should abandon lunar exploration after only six manned landings. They also outline the progress other countries have made in launching missions to the moon. Entries in the final section, "Greenlighting the Red Planet: Should We Go Straight to Mars?" consider whether reaching the fourth planet from the sun—long a subject of science fiction and scientific inquiry—should be the next major goal of the U.S. space program. Contained in the appendix is the official National Space Policy of the United States, which was released by the Obama administration on June 28, 2010.

In closing, I would like to express my gratitude to the authors who have allowed the H.W. Wilson Company to reprint their articles here. I would also like to thank Joseph Miller and Paul McCaffrey for inviting me to return to The Reference Shelf series and edit a book on one of my favorite topics. Special thanks go to my old compatriots Rich Stein and Ken Partridge for their loving care on this volume, as well as to my wife, Ana Maria Estela, who encouraged me to take on this project despite our having a newborn in the house. This book is dedicated to my daughters, Juliana and Olivia, through whose eyes I see this world anew. They fly me to the moon.

Christopher Mari
June 2011

1

The U.S. Space Program: A Laudable Past, a Muddled Present

Courtesy of NASA

Astronaut Edwin E. "Buzz" Aldrin, lunar module pilot of the first lunar landing mission, poses for a photograph beside the deployed United States flag during an Apollo 11 Extravehicular Activity (EVA) on the lunar surface. The Lunar Module (LM) is on the left, and the footprints of the astronauts are clearly visible in the soil of the moon. Astronaut Neil A. Armstrong, commander, took this picture on July 20, 1969.

Courtesy of NASA

Rising on twin columns of fire and creating rolling clouds of smoke and steam, space shuttle *Discovery* lifts off Launch Pad 39A at NASA's Kennedy Space Center in Florida.

Editor's Introduction

The U.S. space program is very much a product of the Cold War. From the conclusion of World War II in 1945 to the collapse of the Communist states of Eastern Europe between 1989 and 1991, the United States and the Union of Soviet Socialist Republics (USSR) competed politically, militarily, and economically in order to assert the supremacy of their respective political and economic systems. The Cold War wasn't merely a terrestrial conflict. By the 1950s, the battle between capitalism and communism had propelled mankind into the cosmos, in what became known as the first "Space Race."

The Soviets won many of the early "battles" in space. On October 4, 1957, they put the first satellite, *Sputnik 1*, into orbit, and on April 12, 1961, they went a step further by making cosmonaut Yuri Gagarin the first human to reach outer space. Responding to the "Sputnik Crisis," a period in which American policymakers worried the nation was falling technologically behind the Soviet Union, President Dwight D. Eisenhower established NASA in 1958. On May 25, 1961, newly elected president John F. Kennedy gave NASA a specific goal, declaring in a speech before a joint session of Congress that the United States would put a man on the moon by the end of the decade. These were bold words, especially when one considers that America had only recently put its first man in space, launching astronaut Alan Shepard a few weeks after Gagarin.

And yet NASA met Kennedy's challenge, landing astronauts Neil Armstrong and Buzz Aldrin on the moon on July 20, 1969. The mission marked the first time humans had walked on the surface of another world. Over the next three years, NASA completed five additional lunar landings, each more daring than the one before. Even the aborted Apollo 13 mission was considered a "successful failure," since NASA engineers troubleshot a severely crippled ship and brought the three-man crew safely back to Earth. During this heady period, it seemed as if the nation was entering into a bold new era of space exploration, laying the groundwork for American astronauts to visit more distant worlds.

As mentioned in the preface, this was not to be. For a variety of economic, political, and social reasons, manned expeditions to the far reaches of the solar system would remain in the realm of science fiction. As far as U.S. policymakers were concerned, the moon landings had signaled victory in the Space Race and confirmed the superiority of American ingenuity. There was no reason to keep spending money to prove that point. As a result, the U.S. manned program shifted its focus to other priorities, such as building the space shuttle and possibly an orbital space station. Plans to return to the moon and perhaps forge ahead to Mars were put on the backburner, where they've remained ever since.

During this same era—the 1970s and 1980s—NASA's unmanned missions began providing earthbound scientists with valuable information on nearly every planet in our solar system. Flyby probes such as *Voyager 1* and *Voyager 2* journeyed to planets like Jupiter, Saturn, Uranus, and Neptune and provided insights the previous generation of astronomers could have only imagined. Later, a number of Mars survey missions—most notably the Mars Pathfinder mission of the mid-to-late 1990s and the ongoing Mars Exploration Rover missions—vastly increased our understanding of the Red Planet. These missions and many others like them have demonstrated that NASA is more than a Cold War relic. It's an agency that has conducted key scientific inquiries for more than half a century, and it remains a vital part of American astronomical research.

The entries in this chapter provide readers with an overview of the U.S. space agency and the challenges it faces. In the first article, "Above and Beyond," André Balogh presents a history of NASA up until 2009, the 40th anniversary of the first lunar landing. In the next entry, "50 and Counting," the National Space Society (NSS) provides a timeline of NASA's major achievements and failures. Sharon Gaudin describes the enormous influence of the Apollo program on modern technology in the chapter's subsequent piece, "Apollo R&D Changed Technology History." The section continues with Andrew Potter's "Fear of a Red Planet Is Just What We Need," a critical look at NASA.

In the next selection, "Delusions of Space Enthusiasts: Sometimes Innovation Gets Interrupted," Neil deGrasse Tyson, director of the Hayden Planetarium in New York City, explains why rocket technology has been frozen in time since the development of the Saturn V rocket that took astronauts to the moon. Finally, Kenneth Chang explores NASA's current quandary—where to go and what to do next in human spaceflight—in his *New York Times* piece, "For NASA, Longest Countdown Awaits."

Above and Beyond*

By André Balogh
History Today, July 2009

It is 40 years since Neil Armstrong took his 'giant leap for Mankind' on the early summer morning of July 20th, 1969. It was the high point of a vast and expensive space programme initiated by President John F. Kennedy in the early 1960s which ended when Apollo 17's lunar module lifted off from the Moon on December 14th, 1972. In just under three and a half years, 12 US astronauts walked on the Moon, drove around in their Moon buggy and thrilled television viewers around the world with their barely believable pantomime on a celestial body 236,000 miles from Earth.

The end came suddenly and space has not captured the public's attention in the same way since, except, in a very different way, in response to the tragedies of the space shuttles *Challenger* in 1986 and *Columbia* in 2003. The Apollo programme had to compete for attention with other major events: the large-scale unrest in the US over civil rights and against the Vietnam war; then, less than a year after the last Apollo mission, the Watergate scandal which brought down President Nixon. Throughout these upheavals, astronauts walked on the Moon, flew the American flag and displayed the might of US technology and resources to massive global audiences in what remains, arguably, the greatest technical achievement of mankind.

The Apollo programme was a child of the Cold War. The technological stakes had been raised by the Soviet Union's launch of Sputnik 1, the first Earth-orbiting artificial satellite, in October 1957. Apollo was the response of a technically fast-developing and confident nation with bewildering reserves of money and talent. It was also symbolic of a different mentality, optimistic, can-do and willing to confront the most awe-inspiring challenges. As Kennedy proclaimed: 'We choose to go to the Moon in this decade and do the other things, not because they are easy, but because they are hard.'

Soviet firsts in space sustained panic in the US over a perceived 'missile gap' and the launch of Sputnik ignited a space race which saw the two superpowers building and attempting to launch a variety of rockets that could double as Intercontinental Ballistic Missiles (ICBMs). We say attempting, because in those early years rockets were less than reliable, with spectacular and very public failures on the launch pads or shortly afterwards at what became Cape Kennedy in Florida (it reverted to its orginal name Cape Canaveral in 1973). It is now known that the early Soviet rocket programme was also punctuated by failures but at that time such setbacks were concealed by the secrecy surrounding the Baikonur space complex in what is now Kazakhstan.

Yet Soviet scientists were soon confident enough to launch Yuri Gagarin into space for a single orbit of the Earth in his Vostok capsule on April 12th, 1961. It marked a new phase in the space race, demonstrating not only immense confidence but also the powerful appeal to the public of men in space. While NASA (National Aeronautics and Space Administration), the then young and vigorous agency of US government in charge of the space effort, prepared to match Gagarin's feat, it was clear that the superiority of the US could only be truly demonstrated by a qualitatively and quantitatively more spectacular achievement. The announcement of the Apollo programme to land a man on the Moon made by President Kennedy on May 25th, 1961, came just six weeks after Gagarin's flight and was intended to neutralise and ultimately trump the Soviet achievement.

Kennedy's speech to a joint session of Congress, always a solemn occasion, stated: 'First, I believe that this nation should commit itself to achieving the goal, before this decade is out, of landing a man on the Moon and returning him safely to the Earth. No single space project in this period will be more impressive to mankind, or more important in the long-range exploration of space; and none will be so difficult or expensive to accomplish.'

This was to be *the* space project of the decade and would not be surpassed by any other country. It was to be a 'giant step' that would convince the world and reassure the American public that the United States, once moved, could do what nobody else could do. This point, the first of Kennedy's proclamation, recognised that the most important and most convincing force in international politics is the perception of superiority. Gagarin's single round-the-world orbit was impressive as it demonstrated the confidence of the Soviet engineers in the reliability of their rocket technology. It certainly impressed the people in the West and it increased the justified fear of the capability of Soviet missiles. This advantage was pushed to its limits the following year by the Soviet leadership, leading to the Cuban missile crisis.

Kennedy's pronouncement that Americans would go to the Moon and back was an altogether different proposition, implying a much greater and broader based technical capability than lay behind Gagarin's flight. The claim to open space for exploration by sending US astronauts to the Moon was the limit of the achievable in manned space exploration.

Kennedy claimed that the goal would be unmatched in its difficulty, an unparalleled technical and human challenge. By the time of his announcement, NASA had been working hard on a broad technical front to build up a credible space programme from scratch in only three years.

This involved a vast and diverse effort, the use of small, suborbital rockets that could take scientific payloads to the edge of space but not into orbit, several medium-sized rockets for Earth-orbiting spacecraft (the best candidates as warhead-carrying missiles) and the large Saturn 1 rocket that was designed under the leadership of Werner von Braun, one of the leaders of Nazi Germany's V2 rocket weapon programme. At the same time, spacecraft started exploring the Earth's own space environment, leading to the discovery of the radiation belts named after James van Allen and the magnetosphere, the magnetic bubble around the planet that protects it from the direct effects of solar wind. These early discoveries were generally made in parallel by both Soviet and American scientists, though the findings of the more open US experts became much better known than those in the pay of the secretive Soviet regime.

NASA became focused on finding the technical solutions to the challenge of taking astronauts to the Moon and returning them safely. The key technical decision concerned the means of carrying them there and back. There were a number of options proposed but the one selected in the end and ultimately carried out involved the use of a vast launcher that did not exist at the beginning of the project, which was to carry all that was needed to the Moon, land a portion of it which would then be able to take off from the Moon and rendezvous with a pod that could carry the astronauts back to a safe splashdown in the sea. This solution presented very serious risks: astronauts would be required to dock in lunar orbit when such a rendezvous had not even been attempted in earth orbit. But those in NASA who proposed it were right to point out that technically this was the only scenario that could be carried out by the deadline.

And so the vast technical might of the United States was put into gear to design, test and carry out the project. A rocket, the Saturn V, was built, based on the Saturn 1A rocket design, a monument to man's technological achievement that when assembled and ready for launch was about the height of St Paul's Cathedral. The Apollo command and service module, as well as the lunar module, were built and tested starting in late 1961 to a very tight schedule. There was tragedy on the way: an accident involving the command module of Apollo 1 during pre-flight tests on January 27th, 1967 killed astronauts Gus Grissom, Ed White (the first US astronaut to walk in space) and Roger Chaffee. Design changes were introduced and by Christmas 1968 Apollo 8 was capable of flying to and around the Moon and back to Earth. Though this flight was made without the proposed lunar landing module, Apollo 8 was able to test many aspects of the ultimate flight scenario. Orbiting the Moon on Christmas Eve, its three astronauts made an emotional live television broadcast of greetings to the Earth.

Apollo 9 was the first flight, in March 1969, of the complete Apollo package in Earth orbit to test all aspects of the final scenario except the Moon landing itself. Then came Apollo 10 in May 1969, a full-scale dress rehearsal. The lunar module was piloted to within 15km of the lunar surface, rejoining the command module in orbit and then returning to the Earth. The two astronauts who got so close were Thomas Stafford, later to command the American crew in the historic Apollo-Soyuz joint test-flight, and Ed Cernan, who later commanded the Apollo 17 flight, becoming the 'last man on the Moon' as he rejoined the lunar module for the final ascent. An iconic image, Earthrise, was taken by the crew of Apollo 10 as it orbited the Moon.

Apollo 11 was launched on July 16th, 1969 with Neil Armstrong, Buzz Aldrin and Michael Collins as the first crew to fly to the Moon and land two astronauts upon it. Armstrong and Aldrin proceeded to plant the American flag into the lunar regolith (the upper layer that consists of rock fragments and dust). Apollo 11 carried a small array of scientific instruments, among them an aluminium sheet used to collect debris from the solar wind. The astronauts then rendezvoused successfully in Moon orbit and returned to a safe splashdown and quarantine on Earth. Kennedy's commitment was fulfilled.

There were in all six successful Moon landings in a way each more successful than the previous ones, with increasingly higher quality television images and the enduring image of the Moon buggy racing across the lunar landscape. Even the flight of Apollo 13, the aborted mission that led to the famous phrase 'Houston, we have a problem', should be remembered for the remarkable technical recovery from a desperate situation, the stuff of Hollywood, which it later became.

While the Apollo programme was wholly motivated by strategic political considerations, the Moon landings also brought important scientific results through observations made on the Moon's surface and by bringing back a total of about 368kg of lunar samples. Scientists began to understand from these the structure and past history of the Moon, almost completely unknown before, by dating the craters and the maria that cover the Moon's surface. The findings of the Apollo mission also support the scenario in which an early version of the Earth was impacted upon by a Mars-sized body that led to enough material being ejected to form the Moon during the dawn of the Solar System.

What about the bill? Kennedy did not put a figure on the cost when announcing the programme (that would have spoilt the effect of the speech) but recognised that no other space project 'will be so . . . expensive to accomplish'. The cost was indeed very high: NASA's budget shows a peak (in 1969 values) of close to $4.5 billion in 1966, with similar figures in the immediately surrounding years, of which about two thirds were devoted to the Apollo programme. Official figures show that the original budget for Apollo was estimated to be about $23 billion and ended up costing between $20 and $25 billion, thus being, seemingly, on budget, a remarkable feat considering its vastness, complexity and novelty. (These costs in current terms would add up to around $150 billion.) Any uncertainty in judging the final costs is due to the abandonment of three further Apollo flights and the difficulty of accounting fully for the expansion in NASA centres when the infrastructure put in place in the 1960s continued to serve the needs of other NASA programmes well beyond Apollo. This financial effort is amazing when the other, simultaneous commitments of the US are taken into account, not least waging a war in south-east Asia with a fully equipped army of almost half-a-million men and women.

The Apollo programme certainly fulfilled one of its principal aims: showing the Soviet Union who was boss in space and, by implication, frontline technology directly transferable to armaments. The relative ease with which the US paid for the programme, as living standards reached heights never seen before, must have impressed the Soviet leadership and their allies. As a weapon of the Cold War, Apollo was a total success, fulfilling its political aim of showing the superiority of the capabilities of the United States to the world. Almost incidentally, the race to the Moon (never seriously joined by the Soviets) was won.

By the time Apollo 17, the final mission, was launched on December 7th, 1972, the importance of space as a battleground in the Cold War was fading. It was generally accepted that, whatever the relative capabilities of missiles with thermonuclear warheads, there was enough weaponry on both sides, East and West, to destroy the world several times over, regardless of who fired first. (The later 'Star Wars' programme initiated by President Reagan in the 1980s was a much less spectacular, primarily technological effort, kept well under wraps by the United States in order to keep ahead in the arms race.) NASA budgets started shrinking; there was a new programme, the Space Shuttle, demanding resources, so Apollo 18, 19 and 20 were cancelled.

One of the Saturn 5 rockets was reused to launch the first ever space station, Skylab, with myriad scientific instruments on board and scientists among its crew. Its achievements were far from negligible but remain largely in the domain of science. One discovery, made by a solar telescope carried on Skylab, revealed that the ebbing and waning of the large dark areas of the Sun's atmosphere, called coronal holes, affected climatic change on Earth.

While the Shuttle programme progressed more slowly and cost more than expected, there was a decline in space activity, with no manned programmes of any sort, let alone anything with the appeal of Apollo. Even the arrival of the Space Shuttle in the early 1980s lacked its excitement. The scientific achievements were more consistent and brought results to a larger number of scientists than Apollo had done, in most part thanks to the European Spacelab module which was carried into space on the Shuttle. Then, on January 28th, 1986 came the *Challenger* accident that made NASA reconsider its strategy and remove the Shuttle from one of its roles, that of launching satellites. The accident also brought to light the fact, obvious to some well before then, that the risks faced by the Space Shuttles were considerably higher than previously advertised to the public and the media and believed even by some of the managers in NASA, if not those with engineering or scientific backgrounds. (This point was tragically reinforced in February 2003 when the Space Shuttle *Columbia* broke up during re-entry into the atmosphere.)

What of recent and current manned space programmes? The long success and occasional drama of the Russian Mir space station over the 15 years of its lifetime, from the first launch in 1986 to the controlled re-entry into the atmosphere in

2001, provided valuable experience of long-duration space missions. The International Space Station (ISS), currently in orbit, is a superior version of the Mir space station, with large laboratories full of expensive equipment exploring the influence of gravity on a wide range of natural phenomena including its effect on blood flow, as well as attempts to grow protein crystals in space, which prompted major advances in our understanding of the structures of complex biomolecules.

But the ISS has taken considerably longer to develop than originally expected. It also suffered from operational restrictions when, after the *Columbia* accident, the Space Shuttle fleet was reduced in numbers and the launch frequency of the remaining three (*Discovery*, *Atlantis* and *Endeavor*) was also cut. Without the help, cooperation and capability of the Russian Soyuz launcher and vehicle it would not have been possible to maintain the ISS. Times have changed indeed since the days of the Cold War, the space race and Apollo.

The Shuttle fleet is due to be retired in 2010, leaving only the ISS, supplied with crew and goods by Soyuz and the newly developed unmanned European cargo ship, the Automated Transfer Vehicle. This restricts its use, in particular because of the restrictions in crew numbers. The maintenance costs of the ISS are very high and, as its usefulness seems to be questionable, this vast investment in space (commensurate with the cost of the Apollo programme) will no longer be exploited. So what comes next?

It seems that the Moon is again on the agenda. Long neglected in favour of the exploration of Mars, a return to the Moon now seems likely in the next decade or two. There are firm plans in the United States: the preliminaries of the programme for a new launcher and new crew vehicle are now financed at a level compatible with a realistic schedule. While both presidents Bush set an eventual goal of landing men on Mars, the more recent initiative envisages a first phase, the return of astronauts to the Moon and even a more-or-less permanently inhabited Moon base. Of course such ambitious plans, due to their length, are at the mercy of changing political priorities and personalities as well as changing global political and economic circumstances. However, it is probably right to be optimistic about a sustained, long-term effort to colonise the Moon in some sense, even if this is not very clear to even the most ardent advocates of manned spaceflight. There are other ambitious nations, naturally the Russians, who have both past experience and capabilities and potentially vast resources, but also the Chinese, the Japanese and the Indians, have all recently sent unmanned scientific spacecraft to the Moon. Possibly the next realistic step will be robotic landers to carry out experiments locally and then to return a larger variety of samples than was available from the Apollo missions.

And then to Mars? This is highly unlikely. Even though no less a figure than Neil Armstrong has been quoted recently as saying that the challenges to land on Mars are not as difficult as the Apollo pioneers faced, a manned mission to Mars seems beyond the reach of affordable technology and political will (not least because of the financial resources that would be required). What was possible in the 1960s to a technically less advanced generation may be out of reach to this and future generations, simply because of the change in attitudes and priorities and the vast increase in the thresholds of risk that are now seen as acceptable. This is one overhead that the Apollo programme did not have to face. Apollo could not be reproduced in the early 21st century, simply because the world we live in is so utterly different from that of only 40 or 50 years ago; sights have lowered as fears have increased. But the Apollo programme happened, men really walked on the Moon and drove their buggies over the lunar landscape and 40 years on we have to recognise the achievement that crowned an undertaking made to that very different world. It remains an inspiration.

ANDRÉ BALOGH *is Emeritus Professor of Space Physics at Imperial College London and Director of the International Space Science Institute in Bern, Switzerland.*

FURTHER READING

John M. Logsdon, *Decision to Go to the Moon: Apollo Project and the National Interest* (Phoenix, 1976); Mark E. Byrnes, *Politics and Space: Image Making by NASA* (Greenwood, 1994); Eugene Cernan and Donald A. Davis, *The Last Man on the Moon and America's Race in Space* (St Martin's Press, 1999); NASA History website http://history.nasa.gov/apollo.html.

50 and Counting*

By Katherine Brick, et al.
Ad Astra, Spring 2008

Five decades after the beginning of orbital spaceflight, 2008 brings a series of major anniversaries—including NASA's 50th. Yet there is much more to accomplish on the high road to space settlement. Your society has been there every step of the way.

NSS [National Space Society] members have witnessed many landmark achievements over the past 50 years. Here are some of the highlights:

1957

SPUTNIK 1 becomes the first man-made satellite of Earth on October 4.

SPUTNIK 2 places the first animal (dog) into orbit on November 3.

1958

First American satellite, EXPLORER 1, successfully launches on January 31 on a modified Jupiter-C rocket, called Juno 1. The rocket is designed by a team led by future National Space Institute [NSI] founder, Wernher von Braun. The scientific payload is designed by Dr. James Van Allen, future NSS governor.

EXPLORER 4 launches July 26 and maps the outer Van Allen belt. The Van Allen radiation belts are named for Dr. Van Allen, their discoverer, on December 6.

NASA is created on October 1.

Future NSS Governor Robert Jastrow becomes the first chairman of NASA'S LUNAR EXPLORATION COMMITTEE, which established the scientific goals for the Apollo landings.

1959

Soviet **LUNA 1** launches January 2 and is the first artificial object to fly past the moon.

PIONEER 4 becomes first American lunar flyby on March 3.

Soviet **LUNA 2** is the first object to impact the moon (September 14), and Luna 3 returns the first images of the far side on October 14.

1960

Frank Drake conducts **PROJECT OZMA**, what many regard as the first SETI [Search for Extraterrestrial Intelligence] project.

America's **PIONEER 5** photographs a solar flare for the first time in March.

The first weather satellite, **TIROS 1**, launches on April 1.

Two **RUSSIAN DOGS** become the first animals returned from orbit on August 19.

NASA'S MARSHALL SPACE FLIGHT CENTER is formed with future NSI President Wernher von Braun as director.

1961

YURI GAGARIN becomes the first man in space and the first to orbit the Earth, aboard Vostok 1, on April 12.

ALAN SHEPARD becomes the first American in space on May 5 during a 15-minute suborbital flight. Shepard was a member of the original NSI board of directors, later becoming an NSS governor.

On May 25, **PRESIDENT KENNEDY** sets the goal of landing astronauts on the moon before the decade is out.

1962

Future NSS Governor **JOHN GLENN** becomes the first American in orbit on February 20.

Congress passes the Communications Satellite Act, creating **COMSAT**.

1963

VALENTINA TERESHKOVA in Vostok 6 becomes the first woman in space on June 16.

1964

RANGER 7 relays first close-range photographs of the moon on July 31.

1965

On March 18, **ALEXEI LEONOV** makes the first space walk. **ED WHITE** makes the first U.S. space walk on June 3.

MARINER 4 returns the first close-range images of Mars on July 14.

On December 4, future NSS Governors **FRANK BORMAN** and **JIM LOVELL** launch on Gemini 7, setting a duration record of fourteen days.

1966

LUNA 9 becomes the first spacecraft to soft-land on the moon on February 3. American **SURVEYOR 1** lands there in June.

Gene Roddenberry creates a new television series called **"STAR TREK."** His wife, Majel Barrett Roddenberry, later becomes an NSS governor and director. Future NSS Governor Nichelle Nichols stars as Lieutenant Uhura.

In November, future NSS leader **BUZZ ALDRIN** sets a record time of five hours thirty minutes of Extravehicular Activity (EVA) during Gemini 12.

1967

Astronauts Ed White, Gus Grissom, and Roger Chaffee are killed when a fire breaks out during a launch pad test on January 27.

On October 10, the **OUTER SPACE TREATY** enters into effect, with nations agreeing not to place weapons in space or claim celestial bodies for themselves.

1968

Future NSS Director and Governor **PETER GLASER** publishes the first paper on the solar power satellite concept.

2001: A SPACE ODYSSEY, co-written by Stanley Kubrick and future NSS Governor Arthur C. Clarke, is released in theaters. NSS Governor Fred Ordway serves as a consultant on the film.

On October 11 the first manned Apollo mission, **APOLLO 7**, launches with Wally Schirra, Walt Cunningham, and Donn Eisele aboard.

On December 21, Bill Anders and future NSS Governors Frank Borman and Jim Lovell become the first astronauts to ride atop a Saturn V rocket. Their **APOLLO 8** mission becomes the first circumlunar flight.

1969

On July 20, **NEIL ARMSTRONG** and future NSS Chairman **BUZZ ALDRIN** become the first humans to walk on the moon.

The Apollo 11 Saturn V booster was managed by NSI Resident Wernher von Braun.

1970

Future NSS Governor **JIM LOVELL** commands the stricken Apollo 13 mission to a safe conclusion.

The first robotic rover, **LUNOKHOD 1**, launches on November 10 and travels the lunar surface for 10 months.

On December 15, Soviet **VENERA 7** is the first probe to soft-land on Venus.

1971

From January 31 to February 9, **APOLLO 14** explores the moon. Future NSI Director and NSS Governor **ALAN SHEPARD** commands the mission.

SOYUZ 11 cosmonauts **GEORGI DOBROVOLSKY**, **VLADISLAV VOLKOV**, and **VIKTOR PATSAYEV** are first to occupy a space station, but die during reentry on June 29.

1972

Apollo program ends with the return of Apollo 17 on December 19. Future NSS Governor **HARRISON SCHMITT** is on board. Schmitt is the first and only scientist to go to the moon.

1973

The first American space station, **SKYLAB**, launches May 14. Skylab's basic concept of reusing existing boosters had been developed by NSI President Wernher von

Braun's Project Horizon in 1959. Future NSS Governor Bill Pogue pilots the final visit to Skylab. Future NSS Governor Gerald Carr commands the mission.

PIONEER 10 provides the first close-up images of Jupiter in December.

1974

THE NATIONAL SPACE ASSOCIATION is founded in June. Wernher von Braun becomes president in August. Von Braun soon becomes chairman, with broadcaster Hugh Downs as president. Chuck Hewitt is the first executive director.

Von Braun remarks, "I know that you are all here . . . because you believe, as I do, that a new organization is needed to communicate the benefits of our national space program to the American public. The National Space Institute, which we are formally launching today, shall perform that function. It is a nonprofit, educational, and scientific organization. The main role of the National Space Institute will be that of a catalyst between the space techonologist and the user. It will attempt to bring to the attention of people the new opportunities offered by advances made in space experiments and space techniques. It will study the feasibility of the application, and the potential uses of space technology as it relates to other human activities."

GERARD O'NEILL writes the seminal article, "The Colonization of Space," for the September 1974 issue of *Physics Today*.

1975

The National Space Association changes its name to the **NATIONAL SPACE INSTITUTE** (NSI).

The **L5 SOCIETY** is founded by Keith and Carolyn Henson to support Gerard O'Neill's ideas for space colonization.

The first **L5** newsletter is published, saying "our dearly stated long-range goal will be to disband the society in a mass meeting at L5." The issue includes a letter of support from presidential candidate **MORRIS UDALL.**

APOLLO-SOYUZ first international docking in space occurs on July 17.

VENERA 9 and **10** send the first pictures from the Venusian surface in October.

1976

VIKING 1 lands on Mars and returns first photos from the surface on July 20. **VIKING 2** lands September 3 and discovers water frost.

1977

Future NSS member Gerard O'Neill publishes ***THE HIGH FRONTIER*** and founds the Space Studies Institute.

STAR WARS opens on May 25 and becomes the most successful movie to date.

1978

VLADIMIR REMEK of Czechoslovakia is the first non-Soviet, non-American astronaut.

The first unmanned **PROGRESS** resupply flight visits **SALYUT 6** on January 22, enabling record long-duration spacefights.

1979

L5's first international chapter is founded in Sydney, Australia.

L5 Society leads successful opposition to the anti-free enterprise **MOON TREATY**, with the guidance of Leigh Ratiner.

NSI receives U.N. recognition.

PIONEER 11 returns the first close-up photos of Saturn.

1980

SALYUT 6 crew **LEONID POPOV** and **VALERY RYUMIN** are the first to spend six months in space.

1981

NASA's first space shuttle, **COLUMBIA**, launches on April 12. L5 member and future NSS Director **MARIANNE DYSON** is one of the first female mission controllers.

1982

The L5 Society's first Conference on Space Development takes place in Los Angeles, California, from April 2–4. Speakers include author **ROBERT HEINLEIN**, astronaut **FRED HAISE**, and future NSS Director **HANS MARK**. This initiates the long-running series of annual conferences, now called the **INTERNATIONAL SPACE DEVELOPMENT CONFERENCES.**

1983

Europe's **SPACELAB** flies on a space shuttle and hosts the first six-person crew.

Longtime NSS member Sally Ride is the first American woman astronaut on **STS-7**, June 18.

1984

Working with NASA, L5 supports the campaign to initiate the **SPACE STATION PROGRAM.**

BRUCE McCANDLESS performs the first untethered spacewalk during STS-41B and the shuttle lands at Kennedy Space Center for the first time.

The first in-space satellite repair and retrieval occurs during **STS-41C**.

Future NSS President and Director **CHARLIE WALKER** is the first industrial payload specialist to fly in space on **STS-41D**, in support of an electrophoresis experiment. This flight also features the test of a new solar array wing.

SVETLANA SAVISTSKAYA performs the first spacewalk by a woman.

1985

An international suite of spacecraft and telescopes study **HALLEY'S COMET**, including the **GIOTTO** spacecraft that takes the first up-close images in early 1986.

Future NSS President and Director **CHARLIE WALKER** flies on two more shuttle flights.

1986

SPACE SHUTTLE CHALLENGER explodes on January 28, killing all seven crew members.

The first module of **MIR** launches on February 20.

NSS member **GERARD O'NEILL** receives the first **ROBERT A. HEINLEIN AWARD.**

1987

The National Space Institute and L5 Society formally merge to become the **NATIONAL SPACE SOCIETY.**

BEN BOVA on behalf of NSI and **GORDON WOODCOCK** on behalf of L5 sign the merger proclamation on March 28.

1988

Backed by heavy support from NSS, Congressman and future NSS Governor **GEORGE BROWN** wins his campaign to pass the **SPACE SETTLEMENT ACT.**

NSS forms a policy committee and conducts its first **SPACE POLICY SURVEY.**

1989

Premier issue of ***AD ASTRA*** magazine is published.

NSS Governor **MARK ALBRECHT** is appointed Director of the National Space Council under President George H. W. Bush.

FRED GREGORY becomes the first African-American to command a space shuttle mission.

1990

Future NSS Director **ROBERT ZUBRIN** gives first public presentation of "Mars Direct" concept at the ISDC in Anaheim, California.

The **MAGELLAN** spacecraft begins mapping Venus.

The **LONG DURATION EXPOSURE FACILITY** shows that seeds can survive vacuum and radiation.

1991

The collapse of the Soviet Union brings dramatic changes to the **RUSSIAN SPACE PROGRAM.**

1992

Future NSS President DAN BRANDENSTEIN commands STS-49. During this mission the crew conducted the initial test flight of Endeavour and performed a record four EVAs to retrieve, repair, and deploy a satellite.

NSS member and STS-45 payload specialist BYRON LICHTENBERG flies an NSS banner into space.

NSS supports a STUDENT PAYLOAD on STS-52.

NSS member and future teacher-in-space, BARBARA MORGAN, receives the first Space Pioneer Award—Space Activist of the Year.

CHICXULUB CRATER is discovered in Mexico, lending support to the theory that an asteroid impact caused the extinction of the dinosaurs.

1993

SPACE STATION FREEDOM, supported by the NSS, survives by one vote in Congress.

DR. ERNST STUHLINGER receives the NSS's first Wernher von Braun award.

The HUBBLE SPACE TELESCOPE is repaired.

1994

COMET SHOEMAKER-LEVY impacts Jupiter in July.

1995

NSS member EILEEN COLLINS becomes the first female shuttle pilot when STS-63 lands in February.

NSS Governor TOM HANKS stars in *APOLLO 13*, the movie.

NSS.ORG is established.

The first Space Shuttle docks with MIR in June, creating world's largest spacecraft.

1996

NSS Director ROBERT ZUBRIN writes *THE CASE FOR MARS* in which he outlines his "Mars direct" plan for a first landing on Mars.

Scientists report that a Mars meteorite may contain FOSSILIZED LIFE, renewing public interest in exploration of the planet.

MIR celebrates 10 years in space.

SHANNON LUCID is the first woman to spend six months in space.

1997

Working with others, NSS wins a campaign to fund the "FRESH LOOK STUDY" of solar power satellites.

Comet HALE-BOPP makes closest approach to Earth. Comet discoverer ALAN HALE collaborates with NSS on future projects.

MARS PATHFINDER lands on July 4 and deploys first rover on Mars.

1998

A record of 13 PEOPLE are in space at one time in January.

NSS Governor Tom Hanks produces "FROM THE EARTH TO THE MOON," an acclaimed mini-series documenting the Apollo program, based on a book by author ANDREW CHAIKIN.

NSS member ALAN BINDER's Lunar Prospector confirms the existence of hydrogen on the moon.

NSS Governor BRUCE BOXLEITNER first stars in "BABYLON 5."

NSS Governor JOHN GLENN flies into space at age 77 onboard STS-95 in October.

In coalition with others, NSS helps win passage of the COMMERCIAL SPACE ACT.

The in-space assembly of the INTERNATIONAL SPACE STATION begins in November.

1999

NSS member EILEEN COLLINS becomes the first woman to command a Space Shuttle, STS-93.

2000

NEAR (Near Earth Asteroid Rendezvous) spacecraft lands on asteroid EROS on February 12.

The first EXPEDITION crew arrives on the International Space Station in November.

2001

After 15 years in orbit, MIR falls into the Pacific Ocean on March 23.

American DENNIS TITO becomes the world's first person to buy his own ticket to space in April.

2002

NSS hosts a space policy forum on SPACE PROPERTY rights at the World Space Congress in Houston, Texas in October.

2003

STS-107 COLUMBIA disintegrates during reentry on February 1. All seven astronauts are killed.

On October 15, SHENZHOU 5 takes YANG LIWEI, China's first astronaut, into orbit.

2004

The Vision for SPACE EXPLORATION is announced by President Bush in January.

The Mars rover Spirit lands in GUSEV CRATER, and Opportunity lands in the MERIDIANI PLANUM of Mars in January.

On October 4, BURT RUTAN, NSS von Braun award winner, wins the $10 million ANSARI X PRIZE with SPACESHIPONE, becoming the first privately-funded, piloted craft to twice enter suborbital space. SS1's engine design is guided by TIM PICKENS, a key member of the HAL5 NSS chapter.

2005

HUYGENS probe lands on TITAN on January 14.

NSS joins successful campaign for NASA to commit to a mission to repair the HUBBLE SPACE TELESCOPE.

2006

NSS Governor and former Director PETE WORDEN takes over NASA AMES RESEARCH CENTER, initiating new efforts in small satellites, participatory exploration, and new partnerships with high-tech companies.

The first privately-owned and financed space station test module, GENESIS I, by BIGELOW AEROSPACE, launches to orbit on July 12.

ANOUSHEH ANSARI becomes the first female fare-paying flight participant.

NASA announces the names of its next generation launch vehicles—Ares 1 and Ares 5—that will send crews and hardware to the moon.

2007

Ashes of former NSS Director and Chairman of the Executive Committee, CHRISTOPHER PANCRATZ, who died in 2003, are launched into space through donations of NSS members.

NSS Governor DR. MICHAEL DEBAKEY is awarded the Congressional Gold Medal.

NSS member BARBARA MORGAN becomes first teacher in space on STS-118 in August.

The National Security Space Office presents a Space-based Solar Power Interim Assessment in October.

Apollo R&D Changed Technology History*

By Sharon Gaudin
Computerworld, July 20–27, 2009

The 40th anniversary of Apollo 11's flight to the moon last week prompted NASA and its supporters to reflect on the technology advances jump-started by the Apollo space program.

The third U.S. space flight program after Mercury and Gemini, Apollo is credited with greatly accelerating the development of several key, still-used technologies, including the integrated circuit, which dramatically altered the face of the computer industry in the 1960s and beyond.

NASA says that other technologies developed at least in part for the Apollo program are now used in products ranging from kidney dialysis machines to water purification systems and athletic shoes.

Experts also noted that without the technology research and development that accompanied the Apollo space missions, top tech companies like Intel Corp. may not have been founded, and we likely wouldn't be using devices like laptops and BlackBerries to post information on social networks like Facebook or Twitter.

The Apollo program was launched in 1961. It started with the ill-fated, never-flown Apollo 1 spacecraft in 1967 and ended with the Apollo 17 mission, which brought the program's final crew to the moon in 1972.

"During the mid- to late 1960s, when Apollo was being designed and built, there was significant [technology] advancement," said Scott Hubbard, a Stanford University professor and a former director of NASA's Ames Research Center in Sunnyvale, Calif.

"Power consumption, mass, volume, data rate—all the things that were important to making space flight feasible led to major changes in technology," he added.

BEYOND THE MOON

Dan Olds, an analyst at Gabriel Consulting Group Inc., said the critical need for better, lighter technology for the Apollo missions, coupled with NASA's financial muscle, pushed advances much further and faster than the fledgling computer industry could have on its own.

"The big problem with inventing and producing breakthrough technology like integrated circuits is the massive cost," Olds added.

"At the early stages of development, everything involved, from tools to production machinery, is a one-off—totally unique and untried," Olds said. "The spending associated with the Apollo missions gave the companies involved the ability to both invent and produce a working chip that made the missions possible."

The program cost $150 billion in current dollars.

Much of the early work on the integrated circuit, the forebear to the microchip, was done under contracts with NASA and the Department of Defense by companies like Texas Instruments Inc. and predecessors to Fairchild Semiconductor International Inc.

"The co-investment between defense and civilian space was very real and hugely important," Hubbard said.

Since the 1960s, the integrated circuit has been a critical piece of many epochal products—"from cell phones to Tickle Me Elmo to the Internet," Olds said.

Without the NASA funding, the technology landscape would probably be far different than it is now, Olds noted. "There would [still] be computers, but they'd be so large and expensive that they would only be used for a handful of specialized applications," he added.

Daniel Lockney, the editor of *Spinoff*, an annual NASA publication that reports on the use of the agency's technologies in the private sector, said that software designed to manage complex systems onboard the Apollo capsules is an ancestor of software now found in devices used to read credit cards. He also noted that liquid-cooled garments based on fire-resistant textiles created for Apollo astronauts are used today by race car drivers and firefighters.

Fear of a Red Planet Is Just What We Need*

By Andrew Potter
Maclean's, August 3–10, 2009

The news media reported last week that NASA's robot rover Spirit, stuck in the Martian equivalent of a ditch, is still spinning its wheels in the deep powder like some suburban doofus trying to free his SUV from a snowbank.

NASA scientists have been working hard trying to figure out some way of rocking the space buggy free, and they hope to give this a shot in a few weeks. But in the meantime, the trapped robot explorer serves as a perfect metaphor for humanity's entire extraterrestrial ambitions.

For space keeners, this should be a week of at least mild celebration. After six tries, the space shuttle Endeavour finally made it into orbit, on its mission to complete the construction of a Japanese-designed veranda that will house science experiments outside the pressurized space station. There are more humans in orbit than ever before, including two Canadians. Encouraging, no?

No. The mission comes framed against the attention given to the 40th anniversary of the Apollo 11 mission that saw humans bounce around for the first time on another world. And in light of what Armstrong and Aldrin accomplished, and the era of great exploration that everyone expected would follow, the baker's dozen of astronauts spinning around in low orbit, still caught in the clutches of the earth's gravitational pull, looks pretty pathetic. As Tom Wolfe, the prose-poet of America's quest for the stars, put it in a recent op-ed for the *New York Times*, "If anyone had told me in July 1969 that the sound of Neil Armstrong's small step plus mankind's big one was the shuffle of pallbearers at graveside, I would have averted my eyes and shaken my head in pity."

But here we are, four decades gone, and the spacefaring dreams of humanity are dead and buried. Not only have there been no manned missions to Mars and no permanent moon bases, no human has so much as ventured out of orbit since

1972. It's as if humanity, having learned to swim by being tossed right into the deep end, opted to spend the rest of the time by the pool clutching the edge.

For decades now, the "space program" has amounted to little more than strapping some humans to a tube, sending them roaring thuggishly up through the atmosphere, and—once finally free of the cloying wetness of air—stopping dead, only to whirl about the earth in the name of science. Imagine if Columbus, having brought the Nina, Pinta, and Santa Maria safely back from the new world, spent the rest of his career tacking back and forth in the harbour at Palos, studying seasickness or testing chronometers.

Of course there are loads of excuses for why we've spent the last four decades doing space doughnuts. It's expensive. It's hard. It's slow. It's cold. There's no air. No gravity. And when they aren't crashing, getting lost, forgetting to return phone calls, or getting stuck in space dust, robots can do whatever sciencey things we need done up there.

But we all know the real reason we abandoned space exploration: Communism failed, the Americans won, and history ended. John F. Kennedy did a good enough job wrapping the moon mission in a lot of "for all mankind" hokey-pokey, but that's not the UN flag stuck in the dirt in the Sea of Tranquility. As the Lyndon Johnson character in *The Right Stuff* put it, "I for one do not go to bed at night by the light of a Communist moon."

The space race, and all the hopes and fantasies it inspired, was always a creature of the Cold War, an exercise in superpower one-upmanship. That doesn't mean the ideals it inspired were false or not worth pursuing, only that it is on this field of striving, the prideful struggle for recognition, that courage, honour, and daring find their home.

There is nothing noble or honourable about our ambitions in space these days, no serious pride to be taken in what we're accomplishing. Putting together the space station is dangerous work, but big deal. So is working on an oil rig, and we don't build monuments or sing hymns to oil rig workers.

It would be nice if the Chinese got more aggressive in space, especially if they were to make a serious go at Mars. Perhaps the fear of the red planet becoming a Red planet would help shake the Americans out of their orbital slumber. But it is not America that is the real problem here, nor is it about "the West." It is the honour of all humanity that is on the line.

Because the odds are that some day, eventually, we're going to be visited by an alien civilization. It may be next week, it may be in the year 12009, but over the near-eternity of time this galaxy is surely going to fill up with a buzzing curiosity of life. Intelligent races will rise who will look to the spiral arms of the Milky Way, wonder what's around the next bend, and set out to take a look.

When they get here, what will they find? An intelligent but distracted species fussing with Facebooks and iPods and Xboxes while a great game unfolds over their heads. Indeed we may have missed our window of opportunity to leave earth; with all the developments in information technology, the appeal of moving in outer space fades in comparison to the easy amusements of virtual space.

But the shame of it all. On their way here the aliens will see the Spirit rover, stuck for millennia in the Martian mud. They will look around and see our footprint on the moon, no bigger than a baseball field. And they'll point at us, galactic laughingstocks, the species that looked briefly to the stars and said, "no thanks."

Delusions of Space Enthusiasts*

Sometimes Innovation Gets Interrupted

By Neil deGrasse Tyson
Natural History, November 2006

Human ingenuity seldom fails to improve on the fruits of human invention. Whatever may have dazzled everyone on its debut is almost guaranteed to be superseded and, someday, to look quaint.

In 2000 B.C. a pair of ice skates made of polished animal bone and leather thongs was a transportation breakthrough. In 1610 Galileo's eight-power telescope was an astonishing tool of detection, capable of giving the senators of Venice a sneak peek at hostile ships before they could enter the lagoon. In 1887 the one-horsepower Benz Patent Motorwagen was the first commercially produced car powered by an internal combustion engine. In 1946 the thirty-ton, showroom-size ENIAC, with its 18,000 vacuum tubes and 6,000 manual switches, pioneered electronic computing. Today you can glide across roadways on in-line skates, gaze at images of faraway galaxies brought to you by the Hubble Space Telescope, cruise the autobahn in a 600-horsepower roadster, and carry your three-pound laptop to an outdoor café.

Of course, such advances don't just fall from the sky. Clever people think them up. Problem is, to turn a clever idea into reality, somebody has to write the check. And when market forces shift, those somebodies may lose interest and the checks may stop coming. If computer companies had stopped innovating in 1978, your desk might still sport a hundred-pound IBM 5110. If communications companies had stopped innovating in 1973, you might still be schlepping a two-pound, nine-inch-long cell phone. And if in 1968 the U.S. space industry had stopped developing bigger and better rockets to launch humans beyond the Moon, we'd never have surpassed the Saturn V rocket.

Oops!

Sorry about that. We haven't surpassed the Saturn V. The largest, most powerful rocket ever flown by anybody, ever, the thirty-six-story-tall Saturn V was the first and only rocket to launch people from Earth to someplace else in the universe. It enabled every Apollo mission to the Moon from 1969 through 1972, as well as the 1973 launch of Skylab I, the first U.S. space station.

Inspired in part by the successes of the Saturn V and the momentum of the Apollo program, visionaries of the day foretold a future that never came to be: space habitats, Moon bases, and Mars colonies up and running by the 1990s. But funding for the Saturn V evaporated as the Moon missions wound down. Additional production runs were canceled, the manufacturers' specialized machine tools were destroyed, and skilled personnel had to find work on other projects. Today U.S. engineers can't even build a Saturn V clone.

What cultural forces froze the Saturn V rocket in time and space?

What misconceptions led to the gap between expectation and reality?

Soothsaying tends to come in two flavors: doubt and delirium. It was doubt that led skeptics to declare that the atom would never be split, the sound barrier would never be broken, and people would never want or need computers in their homes. But in the case of the Saturn V rocket, it was delirium that misled futurists into assuming the Saturn V was an auspicious beginning—never considering that it could, instead, be an end.

On December 30, 1900, for its last Sunday paper of the nineteenth century, the Brooklyn *Daily Eagle* published a sixteen-page supplement headlined "THINGS WILL BE SO DIFFERENT A HUNDRED YEARS HENCE." The contributors—business leaders, military men, pastors, politicians, and experts of every persuasion—imagined what housework, poverty, religion, sanitation, and war would be like in the year 2000. They enthused about the potential of electricity and the automobile. There was even a map of the world-to-be, showing an American Federation comprising most of the Western Hemisphere from the lands above the Arctic Circle down to the archipelago of Tierra del Fuego—plus sub-Saharan Africa, the southern half of Australia, and all of New Zealand.

Most of the writers portrayed an expansive future. But not all. George H. Daniels, a man of authority at the New York Central and Hudson River Railroad, peered into his crystal ball and boneheadedly predicted:

> It is scarcely possible that the twentieth century will witness improvements in transportation that will be as great as were those of the nineteenth century.

Elsewhere in his article, Daniels envisioned affordable global tourism and the diffusion of white bread to China and Japan. Yet he simply couldn't imagine what might replace steam as the power source for ground transportation, let alone a vehicle moving through the air. Even though he stood on the doorstep of the twentieth century, this manager of the world's biggest railroad system could not see beyond the automobile, the locomotive, and the steamship.

Three years later, almost to the day, Wilbur and Orville Wright made the first-ever series of powered, controlled, heavier-than-air flights. By 1957 the U.S.S.R. launched the first satellite into Earth orbit. And in 1969 two Americans became the first human beings to walk on the Moon.

Daniels is hardly the only person to have misread the technological future. Even experts who aren't totally deluded can have tunnel vision. On page 13 of the *Eagle's* Sunday supplement, the principal examiner at the U.S. Patent Office, W. W. Townsend, wrote, "The automobile may be the vehicle of the decade, but the air ship is the conveyance of the century." Sounds visionary, until you read further. What he was talking about were blimps and zeppelins. Both Daniels and Townsend, otherwise well-informed citizens of a changing world, were clueless about what tomorrow's technology would bring.

Even the Wrights were guilty of doubt about the future of aviation. In 1901, discouraged by a summer's worth of unsuccessful tests with a glider, Wilbur told Orville it would take another fifty years for someone to fly. Nope: the birth of aviation was just two years away. On the windy, chilly morning of December 17, 1903, starting from a North Carolina sand dune called Kill Devil Hill, Orville was the first to fly the brothers' 600-pound plane through the air. His epochal journey lasted twelve seconds and covered 120 feet—a distance just shy of the wingspan of a Boeing 757.

Judging by what the mathematician, astronomer, and Royal Society gold medalist Simon Newcomb had published just two months earlier, the flights from Kill Devil Hill should never have taken place when they did:

> Quite likely the twentieth century is destined to see the natural forces which will enable us to fly from continent to continent with a speed far exceeding that of the bird.
>
> But when we inquire whether aerial flight is possible in the present state of our knowledge; whether, with such materials as we possess, a combination of steel, cloth and wire can be made which, moved by the power of electricity or steam, shall form a successful flying machine, the outlook may be altogether different.

Some representatives of informed public opinion went even further. *The New York Times* was steeped in doubt just one week before the Wright brothers went aloft in the original *Wright Flyer*. Writing on December 10, 1903—not about the Wrights but about their illustrious and publicly funded competitor, Samuel E. Langley, an astronomer, physicist, and chief administrator of the Smithsonian Institution—the *Times* declared:

> We hope that Professor Langley will not put his substantial greatness as a scientist in further peril by continuing to waste his time, and the money involved, in further airship experiments. Life is short, and he is capable of services to humanity incomparably greater than can be expected to result from trying to fly.

You might think attitudes would have changed as soon as people from several countries had made their first flights. But no. Wilbur Wright wrote in 1909 that no flying machine would ever make the journey from New York to Paris. Richard Burdon Haldane, the British secretary of war, told Parliament in 1909 that even though the airplane might one day be capable of great things, "from the war point of view, it is not so at present." Ferdinand Foch, a highly regarded French military strategist and the supreme commander of the Allied forces near the end of the First World War, opined in 1911 that airplanes were interesting toys but had no military value. Late that same year, near Tripoli, an Italian plane became the first to drop a bomb.

Early attitudes about flight beyond Earth's atmosphere followed a similar trajectory. True, plenty of philosophers, scientists, and sci-fi writers had thought long and hard about outer space. The sixteenth-century philosopher-friar Giordano Bruno proposed that intelligent beings inhabited an infinitude of worlds. The seventeenth-century soldier-writer Savinien de Cyrano de Bergerac portrayed the Moon as a world with forests, violets, and people.

But those writings were fantasies, not blueprints for action. By the early twentieth century, electricity, telephones, automobiles, radios, airplanes, and countless other engineering marvels were all becoming basic features of modern life. So couldn't earthlings build machines capable of space travel? Many people who should have known better said it couldn't be done, even after the successful 1942 test launch of the world's first long-range ballistic missile: Germany's deadly V-2 rocket. Capable of punching through Earth's atmosphere, it was a crucial step toward reaching the Moon.

Richard van der Riet Woolley, the eleventh British Astronomer Royal, is the source of a particularly woolly remark. When he landed in London after a thirty-six-hour flight from Australia, some reporters asked him about space travel. "It's utter bilge," he answered. That was in early 1956. In early 1957 Lee De Forest, a prolific American inventor who helped birth the age of electronics, declared, "Man will never reach the moon, regardless of all future scientific advances." Remember what happened in late 1957? Not just one but two Soviet *Sputniks* entered Earth orbit. The space race had begun.

Whenever someone says an idea is "bilge" (British for "baloney"), you must first ask whether it violates any well-tested laws of physics. If so, the idea is likely to be bilge. If not, the only challenge is to find a clever engineer—and, of course, a committed source of funding.

The day the Soviet Union launched *Sputnik 1*, a chapter of science fiction became science fact, and the future became the present. All of a sudden, futurists went overboard with their enthusiasm. The delusion that technology would advance at lightning speed replaced the delusion that it would barely advance at all. Experts went from having much too little confidence in the pace of technology to having much too much. And the guiltiest people of all were the space enthusiasts.

Commentators became fond of twenty-year intervals, within which some previously inconceivable goal would supposedly be accomplished. On January 6, 1967, in a front-page story, *The Wall Street Journal* announced: "The most ambitious U.S. space endeavor in the years ahead will be the campaign to land men on neighboring Mars. Most experts estimate the task can be accomplished by 1985." The very next month, in its debut issue, *The Futurist* magazine announced that according to long-range forecasts by the RAND Corporation, a pioneer think-tank, there was a 60 percent probability that a manned lunar base would exist by 1986. In *The Book of Predictions*, published in 1980, the rocket pioneer Robert C. Truax forecast that 50,000 people would be living and working in space by the year 2000. When that benchmark year arrived, people were indeed living and working in space. But the tally was not 50,000. It was three: the first crew of the International Space Station.

All those visionaries (and countless others) never really grasped the forces that drive technological progress. In Wilbur and Orville's day, you could tinker your way into major engineering advances. Their first airplane did not require a grant from the National Science Foundation: they funded it through their bicycle business. The brothers constructed the wings and fuselage themselves, with tools they already owned, and got their resourceful bicycle mechanic, Charles E. Taylor, to design and hand-build the engine. The operation was basically two guys and a garage.

Space exploration unfolds on an entirely different scale. The first moonwalkers were two guys, too—Neil Armstrong and Buzz Aldrin—but behind them loomed the force of a mandate from an assassinated president, 10,000 engineers, $100 billion, and a Saturn V rocket.

Notwithstanding the sanitized memories so many of us have of the Apollo era, Americans were not first on the Moon because we're explorers by nature or because our country is committed to the pursuit of knowledge. We got to the Moon first because the United States was out to beat the Soviet Union, to win the Cold War any way we could. John F. Kennedy made that clear when he complained to top NASA officials in November 1962:

> I'm not that interested in space. I think it's good, I think we ought to know about it, we're ready to spend reasonable amounts of money. But we're talking about these fantastic expenditures which wreck our budget and all these other domestic programs and the only justification for it in my opinion to do it in this time or fashion is because we hope to beat them [the Soviet Union] and demonstrate that starting behind, as we did by a couple of years, by God, we passed them.

Like it or not, war (cold or hot) is the most powerful funding driver in the public arsenal. When a country wages war, money flows like floodwaters. Lofty goals—such as curiosity, discovery, exploration, and science—can get you money for modest-size projects, provided they resonate with the political and cultural views of the moment. But big, expensive activities are inherently long term, and require sustained investment that must survive economic fluctuations and changes in the political winds.

In all eras, across time and culture, only three drivers have fulfilled that funding requirement: war, greed, and the celebration of royal or religious power. The Great Wall of China; the pyramids of Egypt; the Gothic cathedrals of Europe; the U.S. interstate highway system; the voyages of Columbus and Cook—nearly every major undertaking owes its existence to one or more of those three drivers. Today, as the power of kings is supplanted by elected governments, and the power of religion is often expressed in non-architectural undertakings, that third driver has lost much of its sway, leaving war and greed to run the show. Sometimes those two drivers work hand in hand, as in the art of profiteering from the art of war. But war itself remains the ultimate and most compelling rationale.

Having been born the same week NASA was founded, I was eleven years old during the voyage of *Apollo 11*, and had already identified the universe as my life's passion. Unlike so many other people who watched Neil Armstrong's first steps on the Moon, I wasn't jubilant. I was simply relieved that someone was finally exploring another world. To me, *Apollo 11* was clearly the beginning of an era.

But I, too, was delirious. The lunar landings continued for three and a half years. Then they stopped. The Apollo program became the end of an era, not the beginning. And as the Moon voyages receded in time and memory, they seemed ever more unreal in the history of human projects.

Unlike the first ice skates or the first airplane or the first desktop computer—artifacts that make us all chuckle when we see them today—the first rocket to the Moon, the 364-foot-tall Saturn V, elicites awe, even reverence. Three Saturn V relics lie in state at the Johnson Space Center in Texas, the Kennedy Space Center in Florida, and the U.S. Space and Rocket Center in Alabama. Streams of worshippers walk the length of each rocket. They touch the mighty rocket nozzles at the base and wonder how something so large could ever have bested Earth's gravity. To transform their awe into chuckles, our country will have to resume the effort to "boldly go where no man has gone before." Only then will the Saturn V look as quaint as every other invention that human ingenuity has paid the compliment of improving upon.

Astrophysicist **NEIL DEGRASSE TYSON** *is the director of the Hayden Planetarium at the American Museum of Natural History. Tyson's latest book,* Death by Black Hole: And Other Cosmic Quandaries—*an anthology of his favorite* Natural History *essays—has just been published by W.W. Norton.*

For NASA, Longest Countdown Awaits*

By Kenneth Chang
The New York Times, January 24, 2011

Where to next? And when?

For NASA, as it attempts to squeeze a workable human spaceflight program into a tight federal budget, the answers appear to be "somewhere" and "not anytime soon."

When the space shuttles are retired this year—and only one flight remains for each of the three—NASA will no longer have its own means for getting American astronauts to space.

What comes next is a muddle.

The program to send astronauts back to the moon, known as Constellation, was canceled last year.

In its place, Congress has asked NASA to build a heavy-lift rocket, one that can go deep into space carrying big loads. But NASA says it cannot possibly build such a rocket with the budget and schedule it has been given.

Another crucial component of NASA's new mission—helping commercial companies develop space taxis for taking astronauts into orbit—is getting less money than the Obama administration requested. Companies like Boeing and SpaceX that are interested in bidding for the work do not yet know whether they can make a profitable venture of it.

When it comes to the future of NASA, "it's hard at this point to speculate," Douglas R. Cooke, associate administrator for NASA's exploration systems mission directorate, said in an interview.

A panel that oversees safety at NASA took note of the uncertainty in its annual report, released this month. "What is NASA's exploration mission?" the members of the Aerospace Safety Advisory Panel asked in their report.

The panel added: "It is not in the nation's best interest to continue functioning in this manner. The Congress, the White House, and NASA must quickly reach a consensus position on the future of the agency and the future of the United States in space."

A nagging worry is that compromises will leave NASA without enough money to accomplish anything, and that—even as billions of dollars are spent—the future destination and schedule of NASA's rockets could turn out to be "nowhere" and "never."

In that case, human spaceflight at NASA would consist just of its work aboard the International Space Station, with the Russians providing the astronaut transportation indefinitely.

"We're on a path with an increasing probability of a bad outcome," said Scott Pace, a former NASA official who now directs the Space Policy Institute at George Washington University.

A NASA study, completed last month, came up with a framework for spaceflight in the two next decades but deferred setting specific destinations, much less timetables for getting there. One of the study's conclusions was that trying to send astronauts to an asteroid by 2025—as President Obama had challenged the agency to do in a speech last April—was "not prudent," because it would be too expensive and narrow.

Instead, the study advocated a "capability-driven framework"—developing elements like spacecraft, propulsion systems and deep-space living quarters that could be used and reused for a variety of exploration missions.

Meanwhile, in Washington, the fight is less of a conflict of grand visions than a squabble over dollars and the design details of a rocket.

Last fall, in passing an authorization act for NASA, which laid out a blueprint for the next three years, Congress called for NASA to start work on the heavy-lift rocket. It also said that the design should be based on available technologies from the existing space shuttles and from Constellation; that the rocket should be ready by the end of 2016, and that NASA could have about $11.5 billion to develop it.

At the time, Senator Bill Nelson, a Florida Democrat who helped shape the NASA blueprint, said, "If we can't do it for that, then we ought to question whether or not we can build a rocket."

The blueprint, signed into law by President Obama in October, gave NASA 90 days to explain how it would build the rocket.

Two weeks ago, the agency told Congress that it had decided on preferred designs for the rocket and the crew capsule for carrying astronauts, but could [not yet] fit them into the schedule and constraints.

"All our models say 'no,'" said Elizabeth Robinson, NASA's chief financial officer, "even models that have generous affordability considerations."

She said NASA was continuing to explore how it might reduce costs.

A couple of days after receiving the report, Senator Nelson said he had talked to the NASA administrator, Maj. Gen. Charles F. Bolden Jr., and "told him he has to follow the law, which requires a new rocket by 2016." He added, "And NASA has to do it within the budget the law requires."

The track record for large aerospace development projects, both inside and outside of NASA, is that they almost always take longer and cost more than initially estimated. If costs for the heavy-lift rocket swell, the project could, as Constellation did, divert money from other parts of NASA.

Thus, many NASA observers wonder how the agency can afford to finance both the heavy-lift rocket and the commercial space taxis, which are supposed to begin flying at about the same time.

"They're setting themselves up again for a long development program whose completion is beyond the horizon," James A. M. Muncy, a space policy consultant, said of the current heavy-lift design. "The question is, what does Congress want more? Do they [. . .] just want to keep the contractors on contract, or do they want the United States to explore space?"

He called the situation at NASA "a train wreck," one "where everyone involved knows it's a train wreck."

Constellation, started in 2005 under the Bush administration, aimed to return to the moon by 2020 and set up a base there in the following years. But Constellation never received as much money as originally promised, which slowed work and raised the overall price tag.

When Barack Obama was running for president, he said he supported the moon goal. But after he took office, he did not show much enthusiasm for it. His request for the 2010 fiscal year did not seek immediate cuts in Constellation but trimmed the projected spending in future years.

The administration also set up a blue-ribbon panel, led by Norman Augustine, a former chief executive of Lockheed Martin, to review the program. The panel found that Constellation could not fit into the projected budget—$100 billion over 10 years—and would need $45 billion more to get back on track. Extending the space station five years beyond 2015 would add another $14 billion, the group concluded.

The panel could not find an alternative that would fit, either. It said that for a meaningful human spaceflight program that would push beyond low-Earth orbit, NASA would need $128 billion—$28 billion more than the administration wanted to spend—over the next decade.

If the country was not willing to spend that much, NASA should be asked to do less, the panel said.

Last February, when unveiling the budget request for fiscal year 2011, the Obama administration said it wanted to cancel Constellation, turn to commercial companies for transportation to low-Earth orbit and invest heavily in research and development on technologies for future deep-space missions.

The Obama budget requested more money for NASA—but for other parts of the agency like robotic science missions and aviation. The proposed allotment for human spaceflight was still at levels that the Augustine committee had said were not workable.

In pushing to cancel Constellation, one Obama administration official after another called it "unexecutable," so expensive that it limped along for years without discernible progress.

"The fact that we poured $9 billion into an unexecutable program really isn't an excuse to pour another $50 billion into it and still not have an executable program," said James Kohlenberger, chief of staff of the White House's Office of Science and Technology Policy, at a news conference last February.

At the same news conference, Lori Garver, NASA's deputy administrator, noted that Constellation, without a budget increase, would not reach the moon until well after the 2020 target. "The Augustine report made it clear that we wouldn't have gotten to beyond low Earth orbit until 2028 and even then would not have the funding to build the lander," she said. But with the new road map, NASA may not get to its destinations any faster. As for the ultimate goal of landing people on Mars, which President Obama said he wanted NASA to accomplish by the mid-2030s, it is [slipping even] further into the future.

2

At a Crossroads: Should NASA Be Overhauled?

Courtesy of NASA Dryden Flight Research Center (NASA-DFRC)

A drag chute slows the space shuttle *Endeavour* after landing on Runway 22 at Edwards, California, to complete the highly successful STS-68 mission dedicated to radar imaging of the Earth's surface as part of NASA's Mission to Planet Earth program. The STS program will end in 2011 after 30 years of active service.

Courtesy of NASA Johnson Space Center: Earth Sciences and Image Analysis (NASA-JSC-ES&IA)

This is one of a series of digital still images of the International Space Station (ISS) recorded by the STS-110 crew members on board the space shuttle *Atlantis* following the undocking of the two spacecraft some 247 statute miles above the North Atlantic.

Editor's Introduction

Democrats and Republicans rarely find common ground—especially when it comes to funding expensive government programs during times of protracted economic uncertainty—but in 2010, lawmakers on both sides of the aisle seemed to agree on one thing: They hated President Barack Obama's plans for the U.S. space program.

Although the Constellation program, developed under Obama's predecessor, George W. Bush, had been riddled with funding issues and technical setbacks, it at least had a specific goal: returning to the moon via Apollo-style capsules and rockets and establishing a permanent base. The program thrilled many politicians, who saw it as a matter of national pride.

The program Obama proposed after canceling Constellation was altogether different. Sure, it would extend the life of the International Space Station until 2020 and devote billions of dollars to improved robotic probes, new unmanned ships, and as-of-yet untested flight techniques like orbital fuel depots. It would even provide a small increase to NASA's funding. But the new plan would also force NASA to stop developing space vehicles and instead rely on private aerospace companies for the next generation of U.S. spacecraft. At the same time, NASA would pay Russia to fly its astronauts to the space station and low Earth orbit (LEO), since the shuttle program would be canceled. In short, NASA would abandon manned spaceflight for the foreseeable future.

Over the years—particularly in the decades since NASA has limited its manned space program to LEO—many American citizens have called for an overhaul of the space agency. Some scientists who believe manned missions are costly and distracting have pushed for more funding of unmanned exploration. Manned space exploration proponents, meanwhile, have sought more money for their efforts, as well as specific targets to shoot for. Still others have questioned whether NASA, in its present form, is necessary, given the wealth of problems here on Earth. But no one seemed to like or expect the kind of overhaul Obama proposed.

The policy even divided retired astronauts. In an open letter to President Obama, Neil Armstrong, Jim Lovell, and Eugene Cernan—the commanders of Apollo 11, Apollo 13, and Apollo 17, respectively—wrote, "For The United States, the leading space faring nation for nearly half a century, to be without carriage to low Earth orbit and with no human exploration capability to go beyond Earth orbit for an indeterminate time into the future, destines our nation to become one of second or even third rate stature. While the President's plan envisages humans traveling away from Earth and perhaps toward Mars at some time in the future, the lack of developed rockets and spacecraft will assure that ability will not be available for many years." However, Buzz Aldrin, Armstrong's fellow crewman on Apollo 11, disagreed, writing, "I continue to be excited about the development of commercial capabilities to send humans into low earth orbit and what this could ultimately mean in terms of allowing others to experience the transformative power of spaceflight. . . . I applaud the President for his boldness and commitment in working to make this worthwhile dream a reality."

The entries in the chapter weigh the risks and benefits of President Obama's plan. In the first article, "The Wealth of Constellations: Can the Free Market Save the Space Program?" Charles Homans presents an expansive overview of the decision to table Project Constellation. In "Space to Thrive," the subsequent piece, a writer for *The Economist* cheers the president's policy, calling it a needed shakeup for the space agency. In the next article, "Wrecking NASA," aerospace engineer and Mars Society president Robert Zubrin lambastes the president and the thinking behind his decision. In a story for the *Huffington Post*, Rick Tumlinson explores "The Role of Government in a New American Space Agenda," sharing his own thoughts on the subject.

In "Saving Our Space Program," the next entry, Bob Deutsch considers what the revamped space program will mean to the American psyche. The chapter continues with "NASA, We've Got a Problem. But It Can Be Fixed," in which John Tierney considers the role private aerospace companies could play in revolutionizing space exploration, provided President Obama gives them a specific destination, such as Mars. Finally, in a piece for *The Space Review*, Lou Friedman offers his thoughts on how to bridge the longstanding divide between science and manned spaceflight at NASA.

The Wealth of Constellations*

Can the Free Market Save the Space Program?

By Charles Homans
Washington Monthly, May/June 2010

Marine Major General Charles F. Bolden has made a career of taking on daunting assignments. After growing up black in segregated South Carolina, Bolden spent his teenage years badgering congressmen into helping him gain admittance to the nearly all-white U.S. Naval Academy in Annapolis. Shipped out to Southeast Asia in 1972, he flew more than a hundred missions over Vietnam and Laos. After the war he spent a few years test-flying experimental aircraft, then—why not?—became an astronaut. When the space shuttle *Columbia* blasted off from Cape Canaveral in January 1986, sixteen days before the *Challenger* explosion, Bolden was in the pilot's seat. So when President Barack Obama was looking to fill the top job at the National Aeronautics and Space Administration last May, the ex-astronaut, then sixty-two and retired, seemed a natural choice. NASA was four years into its most ambitious project since Apollo, a plan to send American astronauts back to the moon and, in time, on to Mars. It needed a leader equal to the challenge.

Nine months later, however, Bolden received an assignment even tougher than overseeing the mission: getting rid of it.

Under the budget released by the Obama administration in February, NASA was to get out of the business of human spaceflight altogether, at least for the near future—no more space shuttle or rockets, no capsules or moon-landing apparatus. In their place, NASA would oversee something very different: a $6 billion, five-year contract for a handful of private companies to ferry American astronauts to and from the International Space Station—to operate a fleet of space taxis, more or less. Human spaceflight, the province of national identity and aspiration since Yuri Gagarin first hurtled into orbit, was going to be outsourced.

So it was that Bolden found himself, on a Wednesday in late February, sitting alone at a witness table on Capitol Hill—in the same Senate hearing room where he had been lavished with praise during his confirmation the previous summer—facing a panel of livid senators. "I absolutely believe," Louisiana Republican David Vitter declared, "that this budget and the vision it represents would end our human spaceflight program as we know it, and would surrender—at least for our lifetime, perhaps forever—our world leadership in the area." Others railed about the more parochial concerns that accompany the cancellation of giant government programs. "Those seven thousand folks directly in Florida and maybe fourteen thousand others who are impacted—what are we going to say to them?" demanded Senator George LeMieux, Republican of Florida, whose Space Coast depended on NASA and its contractors for jobs. Even Robert "Hoot" Gibson, the commander on Bolden's first shuttle mission, was called to testify against the plan. "This abrupt change in NASA's exploration approach has no clear path," he told the panel. "No destination. No milestones. No program focus."

Bolden's wiry five-foot-seven frame, already diminished by a big-around-the-shoulders suit jacket, seemed to shrink further as the speeches and interrogations wore on—they weren't so much questions to be answered as questions to be endured, and the administrator gamely endured them for an hour and a half. "I—I really do look forward to continuing to work with you all," he said as he finally stood up to leave. "We'll get it right."

In the hallway a scrum of half a dozen reporters was waiting. "Mr. Administrator—"one of them called out, but Bolden just nodded politely and kept walking, the soles of his shoes clopping against the marble. The scrum followed, shouting questions: What exactly was the plan? Did NASA still intend to go to Mars? When? How? The administrator's brisk walk now bordered on a jog. Just before the pack spilled into the rotunda at the end of the hall, an Associated Press correspondent managed to plant himself in front of Bolden. "Mr. Administrator," he said, panting, "what is your vision? *When are you going to have a vision?*" Bolden stopped and looked up at him, a plaintive note in his voice. "I think I *have* a vision," he said.

The Vision: It is constantly invoked at NASA, at once the government's most relentlessly technical and most airily mystical agency. (The Office of Housing and Urban Development does not talk about the Vision.) NASA's goals and ambitions have never been particularly justifiable in any concrete sense; when Wernher von Braun, the ex-Nazi rocket scientist who designed the Apollo program's Saturn rockets, was asked what the purpose of going to the moon was, he replied, "What is the purpose of a newborn baby? We find out in time." You could get away with that in NASA's 1960s glory days, when the agency's accomplishments were splashed across newspapers' front pages and its mantra, posted on the walls of its contractors' rocket factories, was "Waste anything but time."

But once the space race was won and Americans' attentions returned to the planet they lived on, NASA's focus gradually shifted away from the Vision and toward the goal of every aging bureaucracy: survival. The space agency has spent decades in a holding pattern, sinking billions of dollars into projects of questionable usefulness and limited popularity—the space shuttle and the International Space Station—that have kept astronauts and engineers occupied, but have also left the agency even less able to pay for the frontier-expanding ventures that were once its hallmark. One administration after another has struggled to find its way out of this conundrum, announcing Mars missions, lunar bases, and plans for exotic new spacecraft. But these projects have almost always failed to pan out, leaving the agency more cash-strapped than it was before, at which point NASA quietly goes back to the old shuttle routine and hopes that no one really notices. "The agency has been sort of wandering in the desert for forty years," says Scott Horowitz, a former astronaut and Bush-era NASA associate administrator.

What is different now is that for the first time, the old routine is no longer an option. The shuttle is slated to make its final flight next year—the production lines for its fuel tanks have already been shut down—and the agency has nothing ready to take its place. This means that NASA won't have a vehicle to reach the International Space Station it has spent eleven years and $48.5 billion building—and will find its knowledge of how to pull off human spaceflight atrophying rapidly. It's a hell of a bind: NASA must choose between spending a heap of money continuing [to] meet the obligations of the present—the space station missions and other activities in Earth's orbit—and directing its resources toward the ambitions of the future, new exploratory missions that might be decades away from happening and are hardly guaranteed to even happen at all. In the midst of a recession, it can't do both.

The Obama administration's NASA plan is an attempt to escape this fix. While the details have yet to be hammered out, and still await passage by an inevitably hostile Congress, the idea in a nutshell is this: if NASA helps commercial companies get their rockets onto the launch pad, and those companies find a market for their services beyond NASA, the agency's human spaceflight program will finally be free of its expensive obligations to maintain its rudimentary orbit-oriented activities. Instead of spending billions a year on shuttle launches, the agency can simply book astronauts on commercial flights for $20 million or so a pop. With the money it saves, NASA can redouble its research and development efforts to acquire the technology it needs to push the boundaries of exploration once again. It's this potential for the expeditions of the future—and aerospace jobs in the present—that Obama emphasized in a speech at Cape Canaveral in April aimed at stemming a growing political backlash against his new policy. But for this to happen, the other part of the plan has to work—low-earth-orbit space travel has to go from being a resource-sucking government program to an efficient business.

The best reason to think this idea is not completely crazy is because something like it has happened before. In 1925, when air travel was still in its infancy, Congress passed the Kelly Airmail Act, which allowed commercial airlines to bid on U.S. Postal Service delivery contracts. The government backing allowed the fledgling airlines to expand their routes and service offerings; demand for cargo service increased, and, as flight got cheaper and came to be viewed as a normal and non-death-defying means of getting places, passenger service increased, too. Through a combination of private-sector ingenuity and government seed money, a new industry was willed into existence.

But even the earliest, most rickety commercial airlines at least enjoyed the advantage of traveling between places where people lived; it was easy enough to think of things that needed to go from point A to point B, and could stand to get there faster. Commercial spaceflight, by contrast, has to contend with the fact that there aren't many things that actually need to be done in space. In effect, the rocket builders don't just have to figure out how to serve a market—they have to create one. The plan is risky in more serious ways, too: NASA is gambling that private corporations, some of them as yet untested in spaceflight, can carry astronauts to the space station more safely than the space shuttle, even as the agency exercises less safety oversight over them—a plan that runs counter to the recommendations of every major accident investigation NASA has conducted.

That NASA is willing to put all its chips on a bet as dubious seeming as this one is a testament to the uncomfortable straits in which the agency finds itself as it settles into middle age. One of the government's last great throwbacks to the Cold War, NASA's human spaceflight program has finally reached a point where it can no longer ignore the looming question of what it is, exactly, we're doing in space. And it's hoping that its new commercial partners can figure out the answer.

The agency that has landed a dozen men on the moon, photographed the birth of distant stars, and (if you believe the most dramatic of the astronaut-penned op-eds) will find us a way off the only planet our species has ever known once we pollute it out of habitability, operates out of a Washington office space that is more modest than the Bureau of Alcohol, Tobacco and Firearms'. To be fair, the mind-bending stuff mostly happens elsewhere—the sprawling launch complex at Cape Canaveral, the research centers in California and Alabama—but still, if it weren't for the flags, from the outside you could mistake the place for a law firm. One forlorn space suit—Apollo 8 commander Frank Borman's—sits in a display case in the lobby; the security guards have to semiregularly inform perplexed tourists that if they wish to see more of them, the Smithsonian Air & Space Museum is four blocks away.

The headquarters befit a perennially beleaguered organization, one tasked with carrying out what may be the most absurd mandate in the federal government. Every year NASA is asked to do things that border on the impossible, but is given only $18.7 billion to do them—about what we spend annually on farm subsidies, and a rounding error in the $708.3 billion Pentagon budget. The military spends as much on combat-simulation video games each year as NASA spends on the space shuttle.

Still, NASA's budget would probably be adequate for the scientifically path-breaking work that the agency, and nobody else, does: the probes and telescopes charting near and distant corners of the universe, the satellite-based cosmology experiments, the earth observation missions that untangle the complexities of the planet's climate and weather. The problem is that most people don't care about those things. Not long ago, NASA commissioned a study of Americans' attitudes toward the space program by the Center for Cultural Studies & Analysis, a market research think tank that had previously worked for the Six Flags amusement park chain and the National Alliance for Musical Theater. The forty-six-page report the group delivered provides an elegant summation of the agency's bind. "In the mind of the public," it concludes, "human exploration is NASA's brand. The space quest is a human equation, not just a technical mission." The further NASA got from that mission, the more the agency shifted in taxpayers' eyes "from a household item to a luxury item"—the sort of thing that would be nice to have, but not necessarily if money is tight. NASA's two manned ventures, the space shuttle and the International Space Station, didn't really count—what really galvanized Americans were firsts, missions that pushed the boundary of known experience. But those missions were massively expensive, and it was unclear whether the country could muster the will to finance anything like them again. "Failure of NASA's vision to resonate with the American public to the point where it inspires action is a reflection of a larger problem," the analysts write: "the U.S. currently has no larger shared vision into which NASA's vision can fit."

The space agency's last great bid to figure a way out of this spot began on January 14, 2004, with George W. Bush. In a speech at NASA headquarters just shy of the one-year anniversary of *Columbia*'s immolation upon reentry over Texas, Bush laid out a new direction for the agency: he would send astronauts back to the moon for the first time since 1972, he declared, no later than 2020. And the moon was only the beginning—from there they would go to Mars, the grand ambition of space travel since the early twentieth century. "Mankind is drawn to the heavens for the same reason we were once drawn into unknown lands and across the open sea," Bush said. "We choose to explore space because doing so improves our lives and lifts our national spirit." The speech was a fantastically odd moment—Bush had never before evinced much interest in space exploration, and, indeed, the NASA speech was the first and last time he spoke publicly of his vision. The task of filling in the details ultimately fell to Michael Griffin, who took over as NASA administrator the following year.

A brilliant and imperious engineer with half a dozen advanced degrees to his name, Griffin was seen by his admirers as perhaps the closest that NASA could get to another Wernher von Braun, the defining visionary of the Apollo years. There was little doubt that Griffin saw this in himself, too; in his last days in office, NASA spent $57,000 publishing a handsome bound volume of his collected speeches, entitled *Leadership in Space*. If Bush had set the goal, it was Griffin who fleshed out not just the means of attaining it, but also the argument for doing so. In a speech delivered to an industry group in California early in his tenure, he warned that if the United States were to fall behind Russia and China in the quest to expand the frontier of human spaceflight, it would translate into diminished greatness and influence back on earth. "We must think carefully," he said, "about what the world of tomorrow will look like if the United States is not the preeminent spacefaring nation."

Coming from the top of an agency that had grown accustomed to justifying human spaceflight with diplomacy and science, Griffin's blunt nationalism was unabashedly retro—and the plan he had for NASA's next step was no less so. Called Constellation, it consisted of a family of rockets named Ares, a gumdrop-shaped crew module named Orion, and a lunar lander named Altair. Griffin joked that it was "Apollo on steroids," and indeed, the artist's renderings of the components looked like they could have come off von Braun's drafting table. After a quarter century of the space shuttle, here again was the sleek white pillar, twice the height of the shuttle on the launch pad, topped with a nub of a capsule that would carry the astronauts back to the frontiers they had abandoned when "American Pie" still ruled the pop charts.

When he announced his plan, Bush had laid out one hard-and-fast rule: the new mission had to be accomplished within NASA's existing budget, with modest increases to keep pace with inflation. The order suggested just how unserious the president was about the project. Apollo, which aimed to accomplish the same moon shot on a comparable timeline, had at its peak eaten up more than 5 percent of the federal budget, nearly ten times NASA's share in 2004. The space station, a more modest undertaking, was projected to cost the space agency as much as $96 billion over the life of the project; honest outside estimates of a Mars journey pegged its likely cost at five times that.

But in fact, Bush's budget was even more inadequate than it looked at first blush. The official plan was to use the money from the soon-to-be-canceled shuttle program to fuel Constellation's growth. But the administration had ignored the fact that about half the shuttle's budget was tied up in fixed costs that wouldn't disappear with the craft itself, things like test facilities and Mission Control operations. It was magical thinking masquerading as fiscal responsibility.

The short-sightedness of the administration's accounting became clear when Constellation started running into technical problems, as new NASA programs inevitably do. As the Ares rockets began to take shape, engineers realized that the solid-fuel boosters they had adapted from the space shuttle caused a number of problems in the new design. Most critically, they found that when the first stage burned out, the engine could loose violent vibrations up the shaft to the capsule that housed the astronauts, shaking them so badly they would be unable to read the instrument panels. Mitigating the problem would require years of testing, and billions of dollars—which, of course, Constellation didn't have.

Searching for funding within the zero-sum confines of the NASA budget, Griffin began dipping into other projects' accounts. Soon Constellation was not only behind schedule, but forcing the cancellation of more useful NASA research and technology development programs. By 2009 Ares's first flight was optimistically projected for 2015 and realistically years after that, and the estimated cost of building the rockets had grown from $14.4 billion to $35 billion. The moon base, to say nothing of Mars, receded into the distant future. What had begun as an audacious bid for NASA's past glory was quickly becoming a very expensive means of continuing the dull obligations of the present, the perfunctory trips to and from the International Space Station—something Griffin had long scoffed at as a waste of time and resources. And even that wouldn't be possible until years after the space shuttle was scheduled to be retired, meaning that American astronauts who wanted to get to the station would be reduced to paying to hitch rides with the Russians, whose Soyuz rocket and capsule would be the only means of getting into orbit.

Desperate for something to show for their work, Constellation's engineers eventually cobbled together a rocket, the Ares I-X, for an October 2009 test flight. But it was really just a mock-up of the still-nonexistent real thing—assembled from a spare shuttle booster and preexisting military rocket electronics, with a fake upper stage and capsule, it resembled Griffin's original vision in shape alone.

Preoccupied with his ambitions, Griffin had made the classic engineer's error: in drawing up his grand designs, he had simply assumed the money he needed to make them happen would be there. As a result, he left his successors with a far bigger mess than he had inherited. If there was one thing his NASA tenure proved, it was that the human spaceflight program's problem wasn't a lack of ambition—it was getting someone to pay for it. In trying to figure their way out of the same box, his successors would steer a course in the opposite direction.

Although the Obama administration and NASA have remained relentlessly mum about who exactly charted NASA's new course, both its supporters and its detractors tend to believe that its most novel element—the outsourcing of human spaceflight—bears the fingerprints of the agency's new deputy administrator, Lori Garver, rather than Charles Bolden. Garver was Obama's space policy adviser during the campaign, the head of his NASA transition team, and a longstanding champion of commercial space travel. And there's not much on her superior's resume that suggests a desire to upend NASA's status quo—in fact, during the administrator search Bolden had been championed by Constellation supporters like Florida Senator Bill Nelson because he seemed the candidate least likely to scrap it.

Garver landed in space policy more or less accidentally, after working on former astronaut John Glenn's 1984 presidential campaign during her first year out of college. "I definitely fell into it," she told me when I visited her in her Washington office this spring. With her bright lemon-yellow jacket and her chestnut hair pulled into a loose bun, Garver looked less like the archetypal NASA bureaucrat than she did a well-heeled fortysomething suburban lawyer, and she spoke with a conversational ease uncommon in an agency full of engineers and retired military officers. After the Glenn campaign, Garver went on to spend thirteen years at the National Space Society, an advocacy organization aimed at shoring up public and congressional support for space exploration, followed by five years at NASA during the Clinton administration. Her closest brush with spaceflight occurred in 2002, when she was working as a private aerospace consultant. A client of hers, a wealthy CEO and aspiring space tourist, abruptly backed out of his plans to fly to the International Space Station aboard a Russian Soyuz rocket. Desperate to fill the vacancy in the capsule, the Russians dropped the price of the seat, and it occurred to Garver that, with enough sponsorships, she might be able to make the trip herself.

Garver (who had two young sons at the time) dubbed herself the "Astromom," and a deal with the Discovery Channel was soon in the offing. She traveled to the Russian cosmonaut training facility of Star City for medical testing, and went so far as to have her gall bladder removed to meet the physical requirements for the trip. But fate intervened again, and even more bizarrely, in the form of Lance Bass, a member of the boy band N*SYNC, who announced he was also interested in the ticket. Bass and Garver held a surreal joint press conference in Moscow, but by then it was clear that the preteen idol would be able to put together the money for the trip first. Garver returned to Washington, and politics, working on the presidential campaigns of John Kerry in 2004 and both Hillary Clinton and Obama in 2008.

This background—long on politics, minor celebrity, and commerce, short on science, engineering, and government or military service—is unusual for NASA, and it explains part of why Garver has attracted considerable controversy at her new post. During the presidential transition in December 2008, Griffin (who declined to comment for this story) stonewalled Garver, refusing to hand over documentation of the Constellation project to her transition team. When the two bumped into each other at a book party shortly before Christmas, Garver confronted him about it. "Mike, I don't understand what the problem is," she said, according to witnesses who described the incident to the *Orlando Sentinel*. "We are just trying to look under the hood."

"If you are looking under the hood, then you are calling me a liar," Griffin shot back. "Because it means you don't trust what I say is under the hood." (Garver smiled when I asked her about the confrontation. "Transitions are tense times," she said, "and not everyone handles them well.")

When I talked with other veterans of the Griffin era, many of whom professed to have known and liked Garver for years, they spoke of her rise to influence the way a concert violinist would speak of a kid juggling an eighteenth-century Stradivarius next to a fireplace. Their grand fear is that the new NASA plan amounts to the twilight of the heroic engineer: that an agency that seemed poised to reclaim the greatness of its heyday will now be reduced to aimless tinkering. "Constellation got off to a false start, but there was a goal," says Scott Horowitz, the ex-astronaut and NASA official, who worked under Griffin at the agency. "This new budget basically throws out any government ability to provide human spaceflight for the foreseeable future."

Garver, naturally, sees it differently. "This is an attempt to utilize the best the industry has to offer," she told me. "They've all been telling us, 'Government is too hands-on, you're not providing value added.' It's a way for them to lean forward more, us to back off more, and do the harder things." The theory is that the agency can extract itself from the expensive, time-consuming, and not particularly useful business of low-earth-orbit travel—as opposed to long hauls to the moon or beyond—that has defined human spaceflight since the end of Apollo, and free up

NASA's relatively limited resources for the research and development it needs to undertake before it can send astronauts further out into the solar system. Maintaining shuttle service to the International Space Station eats up a third of NASA's annual budget, and for years NASA hasn't really been able to convincingly explain why we're spending the money. The best justification the agency has mustered for the space station—which in truth isn't of much use for scientific research—is that without it, the space shuttle doesn't really have anywhere to go. The main reasoning for continuing the shuttle program, and for coming up with a successor to it, is that otherwise we don't have a means of getting to the space station.

By outsourcing its routine space transportation needs, the agency is essentially throwing the question of what it is we're trying to accomplish by going to space to the private sector. (In April, the Obama administration tried to compromise with Constellation's defenders by proposing keeping the Orion capsule part of the project, but only as an emergency escape vehicle—a kind of lifeboat—for the space station, not a means of transporting astronauts to it.) The ideological contortions involved here attest to just what a strange corner of politics it is that the space program occupies. Republican politicians and conservative pundits have beaten up on Obama for ending one of the most inefficient, government-dominated ventures in history—the *Washington Post*'s Charles Krauthammer called the decision "a constricted inward-looking call to retreat"—in favor of something bordering on galactic Milton Friedmanism: the administration is gambling that in the post-post-Cold War era, the market has a better idea of what's worth doing in orbit than the government does. "It's hard to swallow from a national pride standpoint, but I don't know why it should be," says Marco Caceres, an analyst with the Teal Group, an aerospace market research firm. "The potential is there to grow new industries, to grow our economy. China can put a flag on the moon, but so what? We'll be sending people into low earth orbit, and building whole new businesses. If we let private industry take the lead, it'll be private companies that figure out what's actually worth doing."

The companies the government is trusting to solve this existential riddle are a mix of familiar contractors and would-be gatecrashers, all of which are offering elements of a rocket-and-capsule design which, like Constellation, harkens back to the pre-shuttle space program. One of the two particularly serious contenders is Orbital Sciences, a veteran Pentagon contractor that has been launching military satellites for years. The other is Space Exploration Technologies, or SpaceX, a wildly ambitious start-up launched in 2002 by PayPal cofounder Elon Musk with the intention of building a fleet of commercial rockets that could be launched into orbit at a fraction of the currently available price. Musk is the closest thing the nascent commercial spaceflight industry has to a Howard Hughes, a brilliant polymath with an apparently endless appetite for plowing his own considerable fortune into world-changing endeavors (he is also the founder of the electric carmaker Tesla). Although his company is still in its infancy, SpaceX managed to build a rocket from scratch and loft a satellite into orbit in just six years. That's something that most countries, let alone companies, couldn't pull off.

The biggest hurdle Musk and others have to clear, however, isn't technical. After all, aerospace companies have known how to build rockets for half a century—they just haven't found people to pay for them. Although its plans are still unsettled, NASA is unlikely to need more than a few flights a year to bring astronauts to the International Space Station, and a few more to bring cargo there. A single commercial launch company, however, needs more—possibly many more—launches than that each year to stay in the black. That means it will have to find other customers.

What else could these rockets do besides fly stuff to the space station? The only option that really exists today is satellites. The market for launching them is worth somewhere in the neighborhood of $2.5 billion, but it's barely growing. "It's a very, very conservative market," Caceres says. "It's not one you can say is going to double or triple a few years from now." To grow demand, a start-up would have to offer something radically cheaper than what's available today—for instance, the price of launching a small satellite, Caceres says, would have to fall from the going rate of about $10 million to something on the order of $6 to $8 million—something no aspiring commercial provider can seriously promise.

If the new providers can't offer the bargain-basement prices necessary to expand the satellite market, they'll be stuck trying to undercut the existing players—which isn't much easier. The market is dominated by foreign companies that are subsidized, in fact or in practice, by foreign governments. And because satellites are expensive and there are reliable operators available to send them heavenward, few of their owners are willing to stick with a newcomer through the steep learning curve of its early years—which invariably involve blowing up a lot of rockets—unless it's *really* cheap. (Three of the five Falcon 1 rockets that SpaceX has launched to date have crashed into the ocean—including one that was carrying the ashes of James Doohan, the actor who played Scotty on *Star Trek*.) Sea Launch, a Boeing spinoff that succeeded briefly as a freestanding commercial enterprise, filed for bankruptcy protection last summer, after the cost of its rockets increased and one of them blew up on the launch pad. Customers fled en masse.

Aside from satellites, the other option is the holy grail of the commercial spaceflight industry: flying people to space. But beyond the taxi trips to the space station, the pickings are pretty slim, and mostly theoretical. It's possible that other countries' astronauts would want to buy seats on flights to the space station, but NASA hasn't gauged the extent of the interest. Another option is space tourism. This notion has generated enthusiasm among entrepreneurs; a company called Bigelow Aerospace wants to launch its own inflatable space modules into orbit, which could function as zero-gravity laboratories or even space hotels. And there are inklings of a market here: Richard Branson's Virgin Galactic has done a brisk business selling tickets on future flights aboard the sleek, otherworldly space plane the company has in development.

The problem, though, is that the experience Branson is offering tourists is far cheaper and simpler than the sort of thing that the companies hoping to contract with NASA will be doing with their rockets. Virgin Galactic's aircraft will travel sixty some miles to the edge of the atmosphere and back, what's known as suborbital flight—it technically qualifies as space travel, but it's orders of magnitude less difficult and expensive than reaching orbit, a several-hundred-mile proposition requiring forty times the energy of Branson's day trips. Seats on a Virgin Galactic flight go for $200,000, sailing-around-the-world-on-a-yacht money. But a slot on a commercial flight to the space station would cost *$20 million*. In fact, you can already fly there at that price the way that Lori Garver intended to, with the Russian Federal Space Agency—space tour packages have been available for a decade, and so far there have been only eight takers.

If other customers fail to materialize for the companies that NASA is betting on, the agency finds itself in an awkward spot. The whole point of paying for launch services, rather than bankrolling contractors' ever-metastasizing costs for entire projects, is that it's supposed to be cheaper. But if a company can't find other work, it has to spread its costs out over fewer launches to make its budget—which means the individual launches become more expensive. And if NASA wants to send anything into orbit at all, it has to pay the new prices, whatever they are.

That was how events played out for the Department of Defense fifteen years ago, when it embarked on a strategy similar to NASA's. The military was in the market for a new fleet of rockets to launch defense satellites, to replace its aging Titan IV. At the time, analysts were projecting a huge boom in the demand for commercial satellites, largely driven by an anticipated explosion in the satellite phone market. The Pentagon figured that if it awarded start-up grants to the most promising companies, it could hasten the arrival of a bona fide commercial launch business in the United States; once the rocket builders were up and running, they would have hordes of new customers from the satellite phone industry, and the military could buy their services on a per-launch basis at a reasonable price. Lockheed Martin and McDonnell Douglas (which was acquired by Boeing shortly thereafter) won the contract, and were handed $2.5 billion over the first three years of the project to get the job done.

Over the following years, however, cellular phone companies, the satellite phone providers' rivals in the mobile phone market, began extending the reach of their signals with cheap and easy-to-build repeating towers. The market for cell phones began to grow exponentially, while the one for the far pricier satellite phones all but vanished—along with the demand for satellites. By the time Boeing and Lockheed Martin had actual rockets ready for testing, no one required their services but the military. "The lesson that we learned is that there's market risk, technical risk, and operational risk," says Michael Gass, the president and chief executive officer of United Launch Alliance, the Boeing-Lockheed Martin joint venture that completed the contract. "The companies are good at the technical and operational risks. The market side is something that might be too hard to forecast."

Because the Pentagon needed the rockets, it had no choice but to cover the rocket builders' expenses, which soon grew enormous. The projected cost of a launch jumped 77 percent from 2002 to 2003 alone, and another 29 percent the following year—an increase so great that Defense Secretary Donald Rumsfeld had to formally vouch for the program's necessity to national defense just to make it kosher with defense contracting laws. In 2005 the Defense Department called a spade a spade and quietly redrew the program as the standard cost-plus contracting arrangement that in practice it had already become, and later changed its status to make it subject to fewer budgetary oversight requirements. As of late 2008 the Government Accountability Office calculated its total price tag to date at $8.2 billion—and the rockets were still in the test-launching stage. Even without considering the cost of program development, the price per launch, originally budgeted at a little under $73 million, had grown to $170 million.

If there's a lesson to be drawn from this saga it's that the difference between a government contractor and a commercial company whose only customer is the government is academic at best. And in fact, the Pentagon's experience is already repeating itself at NASA, which in 2006 began a similarly structured $500 million program to commission rockets that could deliver cargo to the space station. SpaceX, the contract recipient that is furthest along in development, was supposed to have tested three Falcon 9 rockets by last September; as of this writing, it was about to test its first. NASA, meanwhile, is asking for an extra $312 million for the companies in its fiscal year 2011 budget to speed things along. And in spite of the risk the agency faces if its launch companies can't find other customers, the agency hasn't commissioned any studies of the potential market for their services—instead, it's relied on the predictions of the companies themselves.

Rocket makers also face a whole new challenge if they want to go from launching satellites and cargo to launching people: namely, not killing them. NASA has said that whatever the next vehicle it entrusts with astronauts' lives is, it should have less than a one in a thousand chance of a fatal accident (the odds on a round-trip space shuttle mission are about one in eighty-five). Commercial providers claim that they can do this without intrusive or expensive NASA oversight. But the fact is that for decades NASA has been experimenting with letting contractors monitor themselves, and the results haven't been good.

The reason for this is that while most rocket-building expertise resides in the private sector—12,000 companies were involved in *Apollo 11* alone—most of the institutional knowledge of how to pull off a manned spaceflight safely remains inside NASA. The less the two overlap, the more crucial information falls through the cracks. Both the *Challenger* and *Columbia* disasters were the result of seemingly minor technical problems whose significance wasn't appreciated up the shuttle program's chain of command. In both cases investigators pinned the blame in part on communication failures between NASA and its contractors, and NASA's diminished oversight over them. Put simply, NASA knew too little about the spacecraft it was flying.

The panel investigating the *Challenger* accident recommended "the total combination of NASA and contractor organizations, working more effectively on a coordinated basis at all levels" in order to avoid repeating the tragedy. At first NASA heeded the investigators' advice, but the inquiry into the subsequent *Columbia* disaster found that once budget pressures reasserted themselves, the shuttle program had gone back to cutting corners. "The change was cost-efficient," the *Columbia* investigators wrote, "but personnel cuts reduced oversight of contractors at the same time that the agency's dependence upon contractor engineering judgment increased." Miscommunication between an oversight-deprived NASA and its contractors has also been blamed for less tragic, more farcical failures. In 1999, the agency lost a $125 million unmanned Mars probe somewhere in the Red Planet's atmosphere due to a navigational failure. The problem, it turned out, was that the contractor, Lockheed Martin, had given NASA all of its data on the program in Imperial units. The space agency used the metric system.

Safely overseeing the development of a commercial venture's human-ready vehicle is possible, of course, but it costs money. How much, exactly, is the multi-billion-dollar question. Would-be taxi providers have thrown out numbers, ranging from a highly unlikely low end of $300 million to an only slightly less unlikely high end of $1.5 billion. But the truth is that no one really knows—NASA hasn't figured out yet what its requirements will be. The agency has spoken of doing less oversight and more "insight"—that is, working with the commercial ventures to come up with a safety code and trusting them to follow it. But it's not hard to see the huge problem that this sets up: would you trust a company whose business is entirely dependent on fixed-price government contracts not to cut corners when nobody's looking?

There's also the question of liability: if something does happen, who's on the hook? Recovering the remnants of *Columbia* and investigating its last flight cost $454 million—enough to sink a fledgling company, which would almost certainly have to put its operations on hold throughout a lengthy investigation. Similarly, a serious problem with the space station could obliterate a company's customer base for months or years. Would-be commercial providers say they wouldn't even get into the business unless the government provided an assurance of bailing them out in the event of a disaster.

All in all, these are not terms that make investors salivate: initial capital in the high hundreds of millions, wildly unpredictable long-term costs, an unproven and possibly nonexistent market, the threat of gargantuan liabilities, and reliance on a single customer that is subject to the whims of Congress and the White House, and might not need the services for more than the next ten years—after all, there's no guarantee that the International Space Station will stay in orbit beyond 2020. Which may be why, for all of NASA's optimism, the people who are willing to sink money into commercial spaceflight are the same as they've always been: existing contractors assured of government backing, and idealistic billionaires. SpaceX, the only serious contender that aspires to be a true freestanding commercial venture, is supported almost entirely by NASA and Musk's own money—although the company is privately held, it lists just two outside investors (including a fund run by Musk's PayPal cofounders), and press reports have pegged their stake in the company at about $35 million, about a tenth of what NASA and Musk are collectively betting on the venture.

In March, while the brouhaha over the new NASA plans was in full swing on Capitol Hill and at Cape Canaveral, a less politically charged bit of news from the agency was mostly lost in the shuffle. Researchers at Cal Tech's Jet Propulsion Laboratory reported that they had successfully uploaded new software to *Opportunity*, one of the two rovers NASA had landed on Mars six years earlier. With the upgrade the rover would be able to use its onboard camera to scan swaths of the Martian landscape in search of rocks of shapes and sizes that suggested scientific interest, and focus on them—to see and think for itself, in a narrow sense.

It was another milestone for one of NASA's most successful endeavors in recent memory. Over the six years they have spent roaming the Red Planet, *Opportunity* and its sibling, *Spirit*, have mined rock samples, charted craters and mountains, and found evidence suggesting the existence of long-evaporated seas on the planet. The rovers have sent back a stream of fascinating photographs: windswept panoramas, microscopic mineral formations, even an eerily moving picture of a sunset, the familiar star appearing from Mars's greater distance as a wan pinpoint of light dropping toward an alien horizon. And they have done all of it for less than a fifth of the budget that the space shuttle program chews up in a single year.

For all the chest-thumping nationalism and science-fiction fantasy that attends human spaceflight, the fact is that most of the truly astonishing things that NASA has accomplished since Apollo have been done not by astronauts in orbit, but by engineers and researchers in labs much closer to home, operating unmanned machines like the Mars rovers and the Hubble telescope. In the years that the shuttle program has burned through upwards of $170 billion—and fourteen crew members' lives—to perform research on subjects like the effects of weightlessness on spiders, earthbound researchers working with a fraction of those resources have discovered hundreds of planets beyond our own solar system, found cosmic evidence of the Big Bang, and begun to pinpoint the age of the Universe. By the time the first astronaut sets foot on Mars—by even the most optimistic estimates—probes and robots will have scoured the planet's atmosphere and surface for half a century.

In turning away from the Bush-era's NASA ambitions, the Obama administration has tacitly acknowledged that human spaceflight, as it has been practiced for half a century, has become something anachronistic and difficult to justify. "We're not trying to rebuild that system," Garver told me. "It was overbuilt." But in betting taxpayer money on the nascent commercial spaceflight industry's promises of an orbiting free market waiting to be born, the agency has shown that it hasn't fully cut itself loose from its old gambler's impulses: the belief that the next paradigm of manned space exploration, the thing that will make all of this worthwhile, is just around the corner. And if not—well, space is a lonely and expensive place.

Space to Thrive*

The Economist, February 3, 2010

In 2004 George Bush announced a plan for America's space agency, NASA, to return to the moon by 2020, land there, explore the surface and set up a base. The moon would then serve as a staging post for a journey to Mars. It was, unfortunately, unclear how this modest proposal would be paid for and, as work began and costs spiralled, the "vision" seemed more science fiction than science.

On February 1st, reality caught up. The back-to-the-moon programme, Constellation, with its Ares rocket, fell victim to Barack Obama's need to find cuts. The Office of Management and Budget described it as over budget, behind schedule and lacking in innovation due to a failure to invest. The office also said Constellation had sucked money from other, more scientific programmes, such as robotic space exploration and Earth observation.

Much has been made of the fact that NASA will, as a consequence of Constellation's cancellation, have to rely on private firms to send its astronauts to the international space station once the space shuttle is withdrawn. In many ways, though, this is the least interesting aspect of what is happening, for what Mr Obama proposed is actually a radical overhaul of the agency.

SUCCESS IS AN OPTION

The rethink looks at four areas: new ways of getting into space; extending the life and use of the space station; the agency's relationship with the private sector; and its scientific mission. The first part of the plan, known as the transformative technology initiative, will cost $7.8 billion over five years. It will develop orbiting fuel depots, rendezvous-and-docking technologies, advanced life-support systems that recycle all of their materials, and better motors for spacecraft. The agency will also develop new engines, propellants and materials as part of a $3.1 billion heavy-lift programme, to allow it to send craft well beyond Earth, while $4.9 billion is allowed for advances in areas such as sensors, communications and robotics.

The second part of the plan is to postpone the death of the space station from 2016 to 2020. More science will be done there (cynics might take issue with the word "more") and there will, specifically, be research into biology, combustion and materials science. There will also be more emphasis on space medicine, and the station is to get a centrifuge. This will allow people to experience artificial gravity in space, which may be important for long-term missions to places such as Mars. Inflatable "space habitats" were mentioned, and these might be used to build extensions to the space station on the cheap. All this will please the station's other participants—Canada, Europe and Japan—which have invested a lot in it for, as yet, little return. It will also help build a coalition of countries that want to travel farther into the solar system.

Now [that] Constellation is cancelled, the plan's third part is to encourage private firms to provide transport to and from the space station. Such journeys into low Earth orbit do not need the heavy-lifting oomph that more wide-ranging missions require, so the proposal is to contract out all of this local delivery work. In fact, such a scheme already exists, and 20 cargo missions by two firms, SpaceX and Orbital Sciences, are planned. The scheme will be extended to include at least two other companies, Boeing and Sierra Nevada Corporation.

Under the new regime, companies will get fixed-price contracts instead of being paid on a "cost plus" basis. The risks and burdens of developing transport to low Earth orbit will thus fall to the private sector. According to Mike Gold of Bigelow Aerospace, a firm that hopes to build inflatable space habitats, such fiscal rectitude has been met with criticism from a surprising quarter: Republican politicians. Bill Posey, who represents Florida in Congress, described it as a "slow death of our nation's human space-flight programme." "If you could fuel a rocket on hypocrisy," Mr Gold suggests, "we'd be on Pluto by now."

The last part of the plan is for more science. The Earth-observation programme will receive some $2 billion to improve the forecasting of climate change and monitor the planet's carbon cycle and its ice sheets. As part of this, NASA will replace the *Orbiting Carbon Observatory*, a satellite that was lost a year ago, and which was supposed to identify the world's sources, and sinks, of carbon dioxide.

There will also be a new emphasis on robotic missions, which are vastly cheaper than manned ones, and cause less angst if they blow up. The first robot destination will be the moon. There will also, according to Charlie Bolden, NASA's administrator, be a mission to the sun, to study the solar wind, and one to improve the agency's ability to detect and catalogue interesting (but potentially dangerous) asteroids that pass near Earth.

It all, then, adds up to a radical shift—but a sensible one after years of fantasy. As Lori Garver, Mr Bolden's deputy, put it, "the old plans lost us the moon. This gives us back the solar system."

Wrecking NASA[*]

By Robert Zubrin
Commentary, June 2010

> We choose to go to the Moon! We choose to go to the Moon in this decade and do the other things, not because they are easy but because they are hard, because that goal will serve to organize and measure the best of our energies and skills, because that challenge is one that we are willing to accept, one we are unwilling to postpone, and one which we intend to win. . . . This is in some measure an act of faith and vision, for we do not know what benefits await us. . . . But space is there and we are going to climb it.
>
> —John F. Kennedy, September 1962

On April 15, Barack Obama traveled to Cape Canaveral. Speaking there to a closed audience of political allies, the president laid out his soaring vision for America's space program. Under the Obama plan, NASA will spend $100 billion on human spaceflight over the next 10 years in order to accomplish nothing.

It must be said that the president phrased his policy wonderfully so that—with the Kennedy Space Center workforce prudently excluded—the camp followers gathered for the occasion had no difficulty in providing the requisite applause. But beneath Mr. Obama's flowery rhetoric, his message was anything but Kennedy-esque. Translated into the English of mortals, he said:

> We choose not to go to the Moon, nor do other things, because they are hard. We do not want a goal that will serve to organize and measure the best of our energies and skills, because that challenge is one that we are unwilling to accept, one we are quite willing to postpone, and one which we will not win . . .

The background to Obama's speech is as follows. In 2004, the Bush administration launched a program called Constellation to develop a set of flight systems, including the Orion crew capsule and the Ares 1 and Ares 5 medium and heavy-lift boosters, that together would allow astronauts to return to the Moon by 2020 and subsequently fly to destinations beyond. Under the plan announced by President Obama, almost all this will be canceled. The only thing preserved out of the past six years and $9 billion worth of effort will be a version of the Orion capsule—but

---

[*] Reprinted from *Commentary*, June 2010, by permission; copyright © 2010 by Commentary, Inc.

one so stripped down that it will be useful only as a lifeboat for bringing astronauts down from the space station, not as a craft capable of providing a ride up to orbit. With the space-shuttle program set to end in the near future, what this means is that the only way Americans will be able to reach even low Earth orbit will be as passengers on Russian launchers.

In his speech, however, the president chose to represent the abandonment of the Moon program not as a retreat but rather as a daring advance. "We've been to the Moon before," he said. "There's a lot more of space to explore." Obama proclaimed it was now time to set our sights on points beyond, to asteroids near Earth, and to Mars. Indeed, he is correct on all counts. But the president's plan makes no provisions for actually following such a course. Instead, it initiates a long stall.

For example, as the first milestone in his allegedly daring program of exploration, Obama called for sending a crew to a near-Earth asteroid by 2025. Such a flight is certainly achievable. All an asteroid mission requires is a launch vehicle such as the Ares 5, a crew capsule (such as the Orion), and a habitation module similar to that employed on the space station. Had Obama not canceled the Ares 5, we could have used it to perform an asteroid mission by 2016—during Obama's own prospective second term. But the president, while calling for such a flight, is scrapping the programs that would make it possible.

The same holds true for the question of reaching Mars. From a technical point of view, we are much closer today to a manned Mars trip than we were to being able to send men to the Moon in 1961, when President Kennedy made his speech committing us to that goal. We reached our destination eight years later. Given true Kennedy-like commitment, we could have astronauts on the Red Planet within a decade. But President Obama chose to set that goal for the 2040s, a timeline so long and hazy as not to require him to actually do anything about it.

Thus, under the Obama plan, NASA will be able to send astronauts anywhere it likes, provided that it begins work toward doing so only after he leaves office.

In an effort to lend the new program some sex appeal, the administration announced, with great fanfare, that its future budgets would provide some funds to support deliveries to the space station by new launch companies. This is a good idea, and long overdue, but not terribly important for the overall future or character of the space program, since NASA has been buying launches from private space firms for the past half-century. A few more launches to low Earth orbit subcontracted out to corporate vendors will change very little.

The man responsible for devising the go-nowhere space policy is the president's top scientific adviser, John Holdren, the director of the Office of Science and Technology Policy (OSTP). According to Holdren, the program's expensive ($10 billion per year) stalling game is justified. Eliminating any focused human-mission goals for NASA will supposedly allow the agency to develop more advanced technologies. This, in turn, will make everything much more achievable at some point in the future, when plans to go somewhere are finally drawn up. To the uninitiated, such arguments may appear plausible, but they are false to the core.

Over the course of its history, NASA has employed two distinct modes of operation. The first, which prevailed in the human spaceflight program during the period from 1961 to 1973, may be called Apollo Mode. The second, prevailing in the human spaceflight effort since 1974, may be called Shuttle Mode.

In Apollo Mode, business is conducted as follows: First, a mission goal is chosen. Next, a plan is developed to achieve this objective. Following this, hardware designs are developed to implement that plan, and, if necessary, technologies are created to enable such hardware. The hardware set is then built, after which the mission is flown.

Shuttle Mode operates altogether differently. In this mode, technologies and hardware elements are developed in accord with the wishes of various technical communities. These projects are then justified by arguments that they might prove useful at some time in the future when grand flight projects are initiated.

Contrasting these two approaches, we see that Apollo Mode is destination-driven, while Shuttle Mode pretends to be technology-driven but is actually constituency-driven. In Apollo Mode, technology is developed to support an overall mission, which means the space agency's efforts are focused and directed. In Shuttle Mode, NASA's efforts are random and entropic.

Imagine two couples, each planning to build their own home. The first couple decide on what kind of house they want, hire an architect to design it in detail, then acquire the appropriate materials to build it. That is Apollo Mode. The second couple poll their neighbors each month for different spare house parts the neighbors would like to sell and buy them all, hoping to accumulate enough material to build a house eventually. When their relatives inquire why they are gathering so much junk, the second couple hire an architect to compose a house design that employs all the knick-knacks they have purchased. The house is never built, but an adequate excuse is generated to justify each purchase, thereby avoiding embarrassment. That is Shuttle Mode.

In today's dollars, NASA's average budget from 1961 to 1973 was about $19 billion per year. That is the same as NASA's current budget. Yet because it had the focus provided by a definite goal that served to "organize its energies and skills," the NASA of the Apollo period was vastly more effective than the equally well-funded agency is today.

Comparing the brilliant record of achievement of NASA's human spaceflight program during the Apollo period with that of the past decade speaks for itself. It also in no way lets administrations between Kennedy's and Obama's off the hook. In technology development, too, the Apollo-era NASA was far superior, creating rocket engines; heavy-lift launch vehicles; space communication, navigation, rendezvous, re-entry, landing, life support, spacesuit, and power technologies, and more—all during a 13-year period. In contrast, during the agency's last quarter century of random research, no new technologies of major significance were developed.

It is this method of constituency-driven, unfocused, never-completed, and perpetually incoherent research activity that Holdren proposes as the basis for NASA's flight into the future.

According to Holdren, the three new "game-changing" technologies that NASA must develop before it attempts to design missions to the asteroids or Mars are electrically powered space thrusters, orbiting depots for propellant storage and refueling, and advanced heavy-lift boosters.

Given current technology, we can do a round-trip mission to a near-Earth asteroid or a one-way transit to Mars in six months—a time no greater than a standard crew shift on the space station. Holdren claims that we need to develop electrically powered space thrusters to speed such trips up. Thus advised, President Obama argued in his April 15 speech that "critical to deep space exploration will be the development of breakthrough propulsion systems." But without gigantic nuclear-powered reactors to provide them with juice, such "breakthrough" thrusters are useless, and the administration has no intention of developing such reactors. So far from enabling quick trips to Mars, the research effort on the unnecessary and unpowerable electric thruster simply provides an excuse for not flying anywhere at all.

The orbital propellant depot's potential utility as a way to enable new manned missions has never been established. To the contrary, none of NASA's recent designs for Moon or Mars missions has involved refueling spacecraft from orbital propellant stations. To insist that mission architects adopt such a strategy because "this is the technology we are working on" is to force the program to accept a suboptimal system design based on an arbitrary decision to favor one technology.

Finally, it is simply not the case that we need new technologies to create heavy-lift launch systems. We not only know how to build them; we actually flew our first heavy-lifter, the Saturn, in 1967, just five years after the Apollo-program contract to create it was signed. In the period since, however, instead of missions requiring booster-production contracts, NASA funded a series of launcher-technology research programs.* None of these resulted in the development of any real-flight hardware. Under the Constellation program, NASA developed a fully satisfactory design for a Saturn 5 equivalent booster, which it called the Ares 5. Yet, instead of proceeding with its development, Holdren has canceled it, promising to produce a new design, after further research, by 2015. But all that is needed to give us a functioning heavy-lift booster is a decision to build it, which will never happen until there is a suitable mission.

Thus, without the guidance supplied by a driving mission, under the new Obama space policy another 10 years and more than $100 billion will be spent by NASA's human spaceflight program without achieving anything significant. We may take part in another 20 flights to low Earth orbit, but together with the Russians, we have already flown there some 300 times over the past half-century. Spending a king's ransom to raise that total to 320 hardly seems worthwhile. Under the Obama plan, we may research some interesting technologies, but without

* Shuttle C, NASP, ALS, NLS, X-33, Spacelifter, and the Space Launch Initiative

a mission plan to guide their selection, they won't be the right technologies, they won't be realized as actual flight systems, they won't fit together, and they won't take us anywhere.

The American people want and deserve a space program that really is going somewhere. President Obama should give it to them. To do that, he needs to put real commitment behind his visionary rhetoric. That means a real program whose effort will commence not in some future administration but rather in his own; one whose goal is not Mars in our dreams but Mars in our time.

> To the frontier the American intellect owes its striking characteristics. That coarseness of strength combined with acuteness and inquisitiveness; that practical, inventive turn of mind, quick to find expedients; that masterful grasp of material things, lacking in the artistic but powerful to effect great ends; that restless, nervous energy; that dominant individualism, working for good and evil, and withal that buoyancy and exuberance that comes from freedom—these are the traits of the frontier, or traits called out elsewhere because of the existence of the frontier.
>
> —Frederick Jackson Turner, *The Significance of the Frontier in American History*, 1893

For many years, some of those on the political left have opposed the space program on the grounds that its funding would do more good if applied to various social needs on Earth. This is, at least, an arguable position.

Yet the administration's proposal to paralyze the space program hasn't a thing to do with competition for funding. Quite the contrary: in order to make the plan palatable to NASA and its contractors, the president has offered to *increase* the agency's funding by $6 billion over the next six years. During the mid-years of the coming decade, for example, Obama's budget proposes to fund the space-station program at a rate of $3 billion per year, even though the nation will neither be conducting any launches to the space station nor building any new modules for it. Billions more will be spent updating the shuttle launch pads, no matter that the shuttles themselves will no longer be flying and that no new launchers will be developed to take their place. Still more will be spent on crew capsules that can only fly down, orbital propellant depots for refueling nonexistent interplanetary spaceships, and electric rockets without sockets to provide them with power.

So if it's not about the money, why does the Obama administration wish to derail NASA? The answer can only be that objections lie not with what the agency *gets* but rather with what it *does*. There is good reason to believe that the administration doesn't like what NASA, and in particular NASA's human spaceflight program, represents.

NASA may be a government agency with the usual bureaucratic attributes, but it is also something else—it is the epitome of the pioneer spirit. The agency's formative adventure—and, in a very real sense, the agency itself—was launched by an administration whose slogan was "The New Frontier." It is not without meaning that so many of its craft have names like Liberty, Freedom, Pioneer, Voyager, Discovery, Endeavor, Pathfinder, Opportunity, and so on. Its astronauts are heroes in the most classic Homeric sense of the term, voluntarily risking death to do great deeds and win eternal glory.

The values championed by the Obama administration are comfort, security, protection, and dependence. But the frontier sings to our souls with different ideals, telling stirring tales of courage, risk, initiative, inventiveness, independence, and self-reliance. Considered as a make-work bureaucracy, NASA may be perfectly acceptable to those currently in power. But for mentalities that would criminalize the failure to buy health insurance, the notion of a government agency that celebrates the pioneer ethos by risking its crews on daring voyages of exploration across vast distances to terra incognita can only be repellent.

There is still a more imperative and even transcendent way in which NASA's human spaceflight program plays within our society's war of ideas. This has to do with our general view of the human future and whether we consider it to be closed or open.

The closed-future theory is one based on the doctrine of limited resources. Its classic formulation can be found in the early-19th-century writings of Thomas Malthus, but in its general form, the construct boils down to this: (A) There isn't enough of x to go around, where x = food supplies, lebensraum, natural resources, carbon-use permissions, etc., as fashion dictates. (B) Therefore, human aspirations must be suppressed. (C) Thus, authorities must be empowered to effect suppression.

The Obama administration has embraced the closed-future theory in its latest "global-warming-requires-carbon-consumption-constraints" incarnation. For Holdren, however, the closed future is not merely a current fad to incorporate into a political portfolio. It has been central to his ideology and preoccupations throughout his career, going back well before global warming became the Malthusian limit du jour. Holdren has co-authored several books with Paul Ehrlich, the Stanford University insect ecologist and perennial panic merchant who wrote the 1968 bestseller *The Population Bomb*. (As a solution to this problem, Ehrlich advocated that the U.S. force sterilization programs on the Third World and set up a domestic Bureau of Population and Environment to issue childbirth permits to American citizens.) In 1988, Holdren and Ehrlich co-organized and led "The Cassandra Conference," which put forth a menu of potential threats usable for justifying global regulation. These included overpopulation, industrial resource exhaustion, acid rain, deforestation, food shortages, energy shortages, the arms race, toxic pollution, runaway technology, and global cooling. In the first chapter of the proceedings of this conference, global "triage" advocate Garrett Hardin writes: "The idea of Progress has become a religion for many in our time. As evidence consider a statement made by the astronaut Scott Carpenter . . . 'I know—I am absolutely positive—that anything a man can imagine, he can accomplish.' "

Which is exactly the fundamental complaint that the closed-future folks have with the human space-exploration program: it makes people think that everything is possible. The issue is not that resources from space might disrupt the would-be regulators' rationing schemes. Rather, it is that the *idea* of an open future with unlimited resources undermines the walls of the mental prison that the self-appointed wardens of mankind seek to construct.

The closed futurists require us to believe that our possibilities are exhausted. But by opening the expansion of the human domain to new worlds, the space program proudly relays the opposite message for all to hear; that we are not done, that far from living at the end of history, we are living at the beginning of history.

It is a message of true audacity and hope, and clearly not acceptable to John Holdren. The question is whether it is acceptable to Barack Obama.

ROBERT ZUBRIN, *an aerospace engineer, is president of the Mars Society and author of* The Case for Mars: The Plan to Settle the Red Planet and Why We Must.

The Role of Government in a New American Space Agenda*

By Rick Tumlinson
Huffington Post, February 16, 2011

In the first decade of this century, the old US space establishment was handed the chance to begin an amazing quest to return to the Moon and explore Mars, *and blew it*—by applying a 20th century state socialist solution to a 21st century frontier challenge, adopting the hardware but ignoring the lessons of Apollo, feeding friends rather than rewarding results, and reinventing the challenge to fit the wheel rather than reinventing itself to fit the challenge.

Meanwhile, a generation of Americans born in the late 20th century who were inspired by and learned the lessons of Apollo invested their funds and ingenuity in what is becoming a new space age. In stark contrast to the managerial mandates of the moribund state program, these new space firms are harnessing the creativity of a new generation, focusing on success rather than spending, applying the principles of a marketbased economy where possible while being willing to pragmatically partner with the state to kickstart their success.

And now the second decade begins. And we have a choice. Either those whose legacy was the creation of the dream and those whose ambition is its realization join together in beginning what may well be the grandest age of human exploration and expansion in history, or the children of Apollo will have to fight their way alone to the gates of the universe while the government space program that inspired and taught them, like some bewildered ancient astronaut, dodders and wanders lost and mumbling into obscurity.

The time for the bold is now, not just on the launch pads of New Space but within the government space community, be it NASA, the Air Force, the Federal Aviation Administration, Congress or the White House—which has so far led a somewhat faltering charge to embrace these new ideas, much to the chagrin of those with vested interest in or clinging to past practices that featured the government as be all and end all of all things space. Yet innovation and ideas are not purely the property of the private citizens; they abound throughout the hallways, cubicles and laboratories of our government, where bright and excited young minds are routinely beaten into boredom by bureaucratic bosses more interested in appearing to toe the line than to think outside of them.

Although it was government that led the first wave over the barricades of gravity, it is now the people who are taking on the work, and all hands are needed—but each in their proper role, applying their own unique tools and working to the common good of all rather than squatting down and fertilizing their own selfdefined tufts of turf.

It's time for a New American Space Agenda, all puns intended. Think of this as part of a manifesto—ideas drawn from almost 30 years of engaging in this revolution, where I have seen firsthand what has and has not worked in an insane battle between the people and their government when it comes to space. It may seem onesided, but frankly it is the government aspect of this endeavor that needs to be forcibly changed, as the market itself as it operates beyond the hallowed halls of Washington, the Pentagon and NASA does a good job of enforcing the rules of the road for the truly private sector in this field. And so I humbly offer five points of purpose and possibility for change in our national approach to opening space.

(If I had a hammer that would allow me to nail these to the national airlock now dividing the people and their government, I would do so.)

I will add more over time for those who are open to listening, but the main points are obvious and simply in need of declaration.

THE NEW AMERICA SPACE AGENDA PART I

The role of government in opening the space frontier is:

1. Declaring the frontier as a place of central national interest where the future ambitions of the people of the nation will be realized in a myriad of ways both tangible and intangible, physical and spiritual, economic, scientific and strategic, and the clear and unambiguous establishment of all necessary fiscal, institutional and regulatory means needed to create, motivate, catalyze and mobilize all the tools, aspects and elements of society, be they directly or indirectly related to the achievement of the goal of creating a free and open frontier in space.
2. Recognizing the frontier as a place where all the grandest aspects of human endeavor are possible and that the advancement of scientific and strategic goals is not simply enhanced but magnified and made purposeful by the employment of the tools of the free enterprise society that allow the government personnel pursuing those goals to do so in the first place.
3. Reaching beyond the realm of private financial investment to explore and support the development of new domains and technologies, always with the goal of passing the fruits of that exploration and development to the peo-

ple—especially those who will make of them new domains for our people and create new wealth for our nation.

4. Providing a cultural framework that supports creativity, innovation, expansion and the application of new ideas and possibilities by educating and inspiring new generations, creating pathways for the realization of their dreams and ambitions and fostering the free flow of ideas between the government, its people and the people of the world in the interest of all.
5. Rewarding those who risk their lives and fortunes to expand the realm of the nation into the frontier by expanding the matrix of governmental supports and services to them, recognizing their rights and the propriety of their property and ideas, enabling and enhancing their abilities to exercise those rights in regards to other nations, and providing the framework of safety and security needed to transform these newly created domains of human activity and ingenuity into places and practices common and conducive to the conduct of normal life of the people of the nation.

We are the children of an earlier frontier. It defines us, it is one thing we do better than anyone else, and one area we are currently sorely neglecting—at exactly the time when it cannot just turn our national economy and malaise around, but if moved from being an oft sidelined sideshow to one of our central organizing national goals could hurl us so far ahead of any others as to assure that our system and way of life will not only shape the future of the planet Earth, but be the model for the future of a new space faring human civilization for thousands of years to come.

If followed and applied, these five points could foster a New Space Age, where this incredible nation of nations can apply all it has to offer in terms of creativity and might in the service of not only assuring its own economic and cultural leadership for eons to come, but leading the people of Earth out and beyond the cradle of this tiny world and into the endless frontier and future offered by the universe around us, a worthy goal not just for the next 10 years, but the next 10,000 and more.

Saving Our Space Program*

By Bob Deutsch
USA Today Magazine, November 2010

Before compromise budgetary legislation was passed earlier this fall, Pres. Barack Obama had called for grounding NASA's space program that would have taken astronauts back to the moon and beyond. In seeking cheaper, faster ways of keeping the U.S. in an exploratory orbit, budgetary issues should not be the only consideration. The country needs a robust space program if we are to realize fully our psychic potential as explorer and player in an ever-expanding frontier.

Our nation finds itself on a treadmill, moving in a simple, two-dimensional motion, burning calories but not exactly sure where it is going. What is required for our national health is a balanced complexity of motion (and its attendant experience) that allows one to feel simultaneously "here" at the center and "out there" at the boundaries. Now, more than ever, in the midst of an economic downturn, and facing a future felt to be receding, Americans need to exist on two planes: the mundane and the mythic.

Our Founding Fathers established the U.S. as an idea, not an ideal. The idea was of liberty and creativity combined, never to be constrained by the status quo. The vision of George Washington, Thomas Jefferson, Benjamin Franklin, and their band of brothers was rooted in the pragmatics of daily life and cast beyond the pale, into the boundless frontier.

The early 21st century certainly is a different time than the late 18th but, underneath the momentary talk, Americans maintain a deep appreciation and need for what this country represents. For instance, in the mid 1990s, I was talking to citizens concerning the proposed luxury car tariff against Japanese automakers, a hot topic at the time. In one discussion, a young woman, a native of Detroit said, "For a couple of generations, members of my family have worked here in the American auto industry. I'm scared about the Japanese competition, but I think tariffs are a bad idea." I asked her why. Upon hearing her response, the earth seemed to shake under the feet of the 11 people in that room, including me. She uttered just three short but profound sentences: "America is a good idea. The idea is freedom. Tariffs are a bad idea." The U.S.'s intrinsic nature is to be open and exploratory. That particularly is a good idea now. Curtailing the space program is a bad idea.

The thinking of this one woman reflects the essence of the relationship that liberty has to creativity—breaking out of routine and expected patterns and going beyond a top-of-mind, business-as-usual, short-term horizon. What is required to live in this frame of mind is having an idea of yourself as one who stands above the press of the moment. Such a mindset allows a practical rootedness in one's authenticity and a "thinking up" that is optimistic and innovative. However, many Americans today feel boxed in by fear. Since words like "ponzi" and "derivatives" have entered their lexicon, they have been living in a question mark—heads down and shoulders hunched in a protective posture.

This is exactly why the U.S. must explore space. We need a program to reimagine a frontier that will allow us to open up this hunkered-down existence. The arguments against it are coldly logical and sometimes all too true. It is too expensive; there is little immediate benefit. The problem with those contentions is that they are blind to the human need to address the cosmic questions of life: Who are we? How are we unique? Why are we here? How did it all begin?

The U.S. has been a place—and should remain one—in which these types of questions are asked. Admittedly, these "big" questions may never be answered satisfactorily but during the search, exploration itself becomes the driving force—our nation in search of the frontier, spatially and experientially. To make this happen, we need a sense of place that includes what is known and what is not, what is possible and what lies beyond our capabilities. Without such urgency of mind, the time frame of our intentions becomes shorter and our motives smaller-minded.

As an anthropologist who traded his backpack and quinine tablets for a Hartmann two-suitor and Dramamine, I have lived among preliterate tribes who have no information technologies, malls, nor media. Their world falls far short of Utopia. Life is hard in the primeval forest but what these weathered-skinned people do have is a general comfort level borne out of an assumed connectedness to their cosmos. Their mythology is rock-solid and enables them to carry on. The U.S. now finds itself "between mythologies." We are not what we once were (mostly because the world changed on us), and we do not yet know what we will become. Where once we had enough resources and weight to overcome any obstacle, we now face a world full of perplexing challenges. Space exploration can help the nation renew its national mythology.

How the country acts during this stage of identity transition will define what it becomes at the conclusion of this national rite de passage. To again take its place as a leader among nations and a beacon for all, the U.S. must remain open, imaginative, and creative; on the intellectual offense; and always exploring. It cannot succumb to the moment's impulse to recoil or resist. Space exploration provides a higher point of view from which to see ourselves. Such a vantage point paves the way for an openness of mind and a generosity of spirit.

We must go beyond metrics and group-think. Data points are not people. Spreadsheets are not artful. To do something artfully requires a dynamic mix of imagination and understanding to see how the world might work. This is not a matter of being correct, but of provoking a self-referring reverie in people that elicits an expanded idea of themselves and their place in the world. As a result, they see anew.

This approach, of course, runs counter to today's government and corporate metric-mania that produces a diminished capacity to conceive bold and innovative visions and strategies. Numbers, budgetary or otherwise, can provide a means for measurement, but cannot "embody" or suggest meaningful insights into the human experience. Yet, such insights are the base coin of national and commercial success.

Creativity calls for a focused subjectivity and the capacity to doubt: an ability to focus on something long enough and deep enough to conjure possibilities not seen in the manifest and immediate moment, along with a healthy acknowledgement that not everything is known. The unknown is fertile soil from which a world of wonders can be cultivated. Here, the plodding of facts and data is circumvented in a nonlinear, symbolic, not wholly rational way. In this maneuver, the mind plays a cognitive trick on itself by creating metaphor. "I will call what I do not know by the name of something that I do know." Suddenly, you become free to explore conceptually. You are released from the rut of the "now" and the already-known. Through this mental leapfrog, the creative impulse extrapolates into unknown scenarios. It moves from the past to instigate an inkling that lays the basis for the beginning of a new narrative, to a springboard that weaves a web of new patterns and associations, to an insinuation of the future as projected in metaphor. This process produces, from the outside-objective point of view, what can be perceived as seemingly off-topic meanderings, but nothing is further from the truth.

What is in operation is a kind of playfulness with ideas that is essential for creativity. This toying around contains a bunch of no's, as in no pretense, analyzing (yet), doubts, pressure to conform, restrictions, and, perhaps most important of all, judgment. Those who are playfully creative possess a curiosity given backbone by their expectation that they will find what they seek even though they do not know what it is they seek (often a statement of fact in space exploration).

In this special state of mental weightlessness, all inhabitant are joined by a belief in a beautiful human quality, directed serendipity: I have a plan because the plan allows me to begin to move forward and, in doing so, I learn about myself. You sort of go down a path and things evolve. By adapting and adjusting to randomness, you shape, but do not control, your endpoint. Yet, you define your endpoint by your own reaction to it—ah, ha! I like this. This is for me. This is me.

However, in sharp contrast, many more Americans today are losing hope in the ties that bind hard work to success. Many see the future as "closing." This mentality foreshortens their vision of self, others, and the world. This orientation, about almost everything, is defensive. Listen to the tone: money makes the world go around; now I have less money and hope—or, I feel better when I see someone worse off than me; I have to fight for everything, and I don't have a lot. In other words, what's the point?

The U.S. was founded on the idea of never accepting the status quo and always exploring further. It is our national heritage, and it is not nice to fool with a nation's nature. NASA's manned space program, particularly in times of uncertainty and fear, can help remind each American what it means to look up and open up—to have an idea of "you" that has a little elbow room.

Only the space program can provide a national effervescence that can give people that boost necessary to investigate their own essence, to write their own story. Manned exploration of space instigates a reverie to help people feel they are more fully alive and participating in a quest beyond the mundane. The very idea of breaking past Earth's pull can help keep people from being inundated by the contingency of any moment.

When thinking about one's own life, there is a sense of freedom in keeping the mind's eye oriented to the "out there, beyond the boundaries" of daily existence. To hold dear this attitude, it helps to realize that astronauts, space stations, the Hubble telescope, etc. represent—and are a reflection of—who we are. More than any other idea, the NASA program allows people to experience the reality that we are but a small speck amidst the immensity of intergalactic space and that we are one with it. To partake daily of this mystery, to wonder, to feel and never flat-line emotionally, lies at the core of the idea of manned space exploration. It is there to help each American—in the context of his or her own life—soar and explore.

BOB DEUTSCH *is a cognitive anthropologist and founder of the firm Brain Sells, a strategic branding and communications consultancy based in Boston, Mass.*

NASA, We've Got a Problem. But It Can Be Fixed*

By John Tierney
The New York Times, April 12, 2010

During his trip to the Kennedy Space Center this week, President Obama will be confronted with an awkward political reality: when it comes to space exploration, sometimes even Republicans can be passionate advocates for the public option.

They've joined with Democratic colleagues in Congress to oppose Mr. Obama's plan to reduce NASA's missions and to encourage private companies' rockets to haul cargo and astronauts into space. Mr. Obama's critics accuse him of abandoning NASA's glorious trailblazing tradition—and maybe, if your constituents have jobs at NASA, it's possible to see something pioneering about an agency whose flagship goes a few hundred miles into space.

But four decades after the moon landing, why isn't NASA venturing somewhere more exotic? The glory days of the Apollo moon missions were possible only because of the race against the Soviet Union, and Mr. Obama could revive that spirit by starting a new space race. But not if he gives in to demands to keep NASA wobbling along in current fashion.

The main problem with NASA is not lack of money. Its current budget is about the same size, when adjusted for inflation, as the average during the 1960s and early 1970s. But space exploration has become so costly that this level of financing won't even pay for a return to the Moon anytime soon, which is what prompted the White House to cancel the Bush administration's lunar mission.

Normally, once a pioneer makes the first trip somewhere, the cost goes down as others follow and technology improves. That's why so many colonists could follow Columbus to the New World, and why the masses today can afford to fly in Lindbergh's path back to Europe. The real costs of shipping freight by rail and air have declined by an order of magnitude since locomotives and airplanes were invented.

In space transportation, though, many costs have actually risen since the days of Apollo. NASA estimates that each seven-person space shuttle mission now costs about $65 million per astronaut, and outside experts say the true cost, once you add in all shuttle-related expenses, is double or triple that figure. Even a cramped trip to the space station on the Russian government's Soyuz will cost the United States more than $50 million per astronaut.

Elon Musk has promised to do the job for just $20 million by building his own rocket, which is being readied for a test launch next month. His company, SpaceX, is saving money through innovations like assembling the rocket horizontally inside a prefab hangar here at Cape Canaveral, thereby avoiding the costs of the customized towers and scaffolding used by NASA to assemble its rockets vertically.

The most important innovation for the space program, Mr. Musk said, would be to get away from the old "cost plus" model in which NASA contractors are guaranteed a profit on top of whatever costs they incur.

"The cost-plus approach encourages aerospace companies to find the most expensive way to do something and drag it out as long as possible," said Mr. Musk, the entrepreneur who co-founded PayPal. "Future contracts should be given to meet milestones based on objective design reviews and actual hardware completion. If a company meets the milestone, they get paid. If not, they don't."

Under Mr. Obama's new plan, there would be further encouragement for SpaceX and other private companies. The traditional aerospace companies remain leery of the plans [. . .] but there are other upstarts besides SpaceX working on projects to get humans into space, and they would be potential competitors.

While private companies would ferry humans and cargo into orbit, NASA would be redirected into basic research and development of technologies for trips far beyond Earth, according to Mr. Obama's plan. This R&D, as White House officials are quick to note in preparation for their Florida trip, would provide jobs in some of the Congressional districts affected by the closing of the shuttle and lunar programs.

In theory, it could make sense for NASA to be doing the high-risk advanced research that could yield breakthroughs for a trip to Mars. But how much progress would NASA actually make?

One way to encourage innovation would be to convert some of NASA's divisions into versions of the Jet Propulsion Laboratory, which has successfully sent robotic probes across the solar system. The laboratory is financed by NASA but operated by CalTech, and therefore freer to innovate outside of political and Civil Service constraints. A presidential commission in 2004 recommended that NASA convert some of its research centers to this academic model, but the agency's bureaucracy showed little interest in surrendering control.

Another way to prod NASA would be to set specific goals. Robert Zubrin, the president of the Mars Society, argues that NASA has been effective only when given a definite destination and hard timetable—the Apollo Mode, as he calls it, which is still used successfully on robotic missions.

But since the moon landings, he said, the human spaceflight program has been stuck in the Shuttle Mode, in which the agency develops technologies to please bureaucratic and political constituencies. Instead of looking for the best ship to reach a goal, it looked for a goal (like the space station) to justify the ship. To Dr. Zubrin, the new Obama plan looks like more of the Shuttle Mode.

"The plan proposes to spend $100 billion on human spaceflight over the next 10 years in order to accomplish nothing," he said. "If we are going to have a NASA human spaceflight program, it must be given a goal."

Dr. Zubrin would like to see NASA committed to a goal like building a ship to reach an asteroid by 2016 (a year that would just happen to fall within a second Obama term) as preparation for a journey to Mars. That's not a bad goal, although I can't see how the agency could afford to make the trip itself.

What it could do is oversee a competition among companies like SpaceX or whatever nonprofit groups wanted to participate. It could offer a series of prizes, like the cash bonus that Lindbergh collected for crossing the Atlantic. Or there could be contracts given to the competitors that would be based strictly on their performance, not on their costs.

What if Mr. Obama, in Kennedyesque fashion, started a race to Mars in which NASA set milestones along the way and rewarded competitors only when they passed each one?

"It'd be cool to set a Mars goal," Mr. Musk said. "Personally, I think it could be done in 10 years." Maybe he's being overoptimistic, but what's the harm in letting him—and everyone else—go for it?

Merging Human Spaceflight and Science at NASA*

By Lou Friedman
The Space Review, February 7, 2011

I really liked what NASA Administrator Charlie Bolden had to say about the news last week that the Kepler mission had discovered a plethora of possible planets around other stars. Some of them are candidates for being Earth-like in size, orbit, and maybe even composition. Bolden said, "In one generation we have gone from extraterrestrial planets being a mainstay of science fiction, to the present, where Kepler has helped turn science fiction into today's reality. These discoveries underscore the importance of NASA's science missions, which consistently increase understanding of our place in the cosmos."

That last sentence captures the huge dichotomy which is NASA. From its very beginning to the present day, NASA provides important, exciting, and popular new discoveries that increase understanding of our place in the cosmos. As such, it remains a symbol of can-do for America and inspiration for the world. Best of all, NASA substantially increases the body of knowledge so important to educating the public, especially schoolchildren, about our planet and our universe.

Unfortunately, there is another side to NASA's story—the human spaceflight program stuck in Earth orbit, mired in politics, and drifting from proposal to proposal, never alighting on one long enough to have a clear purpose. It doesn't have to be this way. For years, I, along with others, have been calling for more integration of science and exploration. With some justification, many science advocates fear such a melding, worrying that integration would mean their projects would be eaten up by the larger human spaceflight program.

That is a legitimate concern if human spaceflight remains without a science or exploration goal. Instead of human spaceflight swallowing science, I'd like to see the reverse: science swallowing human spaceflight by focusing it on exploration. Make exploration more than the name of a program office.

The bollixing up of NASA's program planning by last year's Congress and the emphasis on budget cuts by this year's Congress create a severe challenge for the fu-

* Article by Dr. Louis Friedman originally from *The Space Review*.

ture of human spaceflight. But that challenge also creates an opportunity. Perhaps now is the time to return to that post-Columbia accident debate about the purpose of human spaceflight, to examine what is worth the high cost and high risk of humans in space. I have no doubt that the answer will remain what it has always been when those debates were held: the exploration of other worlds.

Much has been written about shrinking NASA's Apollo legacy infrastructure. That has proved politically impossible as members protect local interests of NASA centers and industry. But there is a shift now, propelled by reduced spending and pressures for reduced government. One possible result is putting a lid on NASA spending and then pushing the lid down to make everything smaller. That would be too bad: goals, missions, accomplishments, and NASA's very purpose would all diminish.

Instead, perhaps we can think about what the public cares about from the space program: scientific discovery, new achievements, and inspiration. Perhaps we can examine what policy makers really want from our space program when they use that vaunted phrase, "American leadership." Doing the same things on a reduced budget doesn't sound like American leadership. Leading other nations in exploration of the universe—and in understanding our own planet and place in the cosmos—does.

If we merge human exploration into science, then admittedly we will reduce some near-term human space program expenses. But that is going to happen anyway. NASA is already being pushed to get out of transportation and focus on exploration. We can build a stronger and more purposeful human program by involving human spaceflight in the programs that are making exciting discoveries about other worlds and our own.

This huge step involves some huge shifts. The biggest shift is the first one: that of the paradigm. Merging human spaceflight mission development into science planning would create enormous program, institutional, and infrastructure upheavals. We have to start from the top, defining our goals and objectives.

The observation, monitoring, and understanding of Earth as a planet is one goal. Another is learning more about near-Earth objects [NEOs], including the discovery, characterization, and use of NEOs, as well as protecting Earth from them. The exploration and possible settlement of Mars is an obvious third goal, while the fourth is the one that Mr. Bolden mentioned, understanding our place in the cosmos. My successor at The Planetary Society, Bill Nye, sums it up by saying we must "know our place in space."

Those goals all have homes in the Science Mission Directorate. Would we dare put the human mission planning in those homes? Many scientists pooh-pooh human spaceflight, and their response might be to cancel it. But most of the scientists involved in space exploration understand that humans are part of that exploration.

Despite the joys of finding extrasolar planets, exploring new canyons and plains on Mars, seeing the edge of the Universe, and learning about our near-Earth environment, it's my view that NASA is in crisis. Its public image is fuzzy and uncertain, and all the political pressures are negative. But despite that crisis, the agency is strong right now: performing missions brilliantly and advancing science and technology. The time to deal with crisis is when you are strong. Now is the time for some new thinking where human spaceflight fits in NASA's future.

LOU FRIEDMAN *recently stepped down after 30 years as Executive Director of The Planetary Society. He continues as Director of the Society's LightSail Program and remains involved in space programs and policy. Before co-founding the Society with Carl Sagan and Bruce Murray, Lou was a Navigation and Mission Analysis Engineer and Manager of Advanced Projects at JPL.*

3

Outer Space Goes Out to Bid: Should Private Companies Lead U.S. Spaceflight?

Courtesy of AFP/Getty Images

This photo, obtained from the Aero-News Network, shows *SpaceShipOne* (bottom craft) riding over the Mojave Desert in California while tethered to the launch ship, the *White Night* (top craft), in its first attempt to win the $10 million Ansari X-Prize, on September 29, 2004. *SpaceShipOne* made its first successful journey into suborbital space in the quest for the X-Prize.

Courtesy of NASA/Kevin O'Connell

SpaceX's Falcon 9 rocket and Dragon spacecraft lift off from Launch Complex-40 at Cape Canaveral Air Force Station, Florida.

Editor's Introduction

In 2010, NASA launched the first phase of its Commercial Crew and Cargo Program, the stated goal of which is "investing financial and technical resources to stimulate efforts within the private sector to develop and demonstrate safe, reliable, and cost-effective space transportation capabilities." Under the terms of the first development phase, five American companies were given a total of $50 million from the American Recovery and Reinvestment Act of 2009—more commonly known as the Recovery Act or "the stimulus"—to develop spacecraft concepts. In October 2010, NASA invited additional proposals from the private sector, and in April 2011, the agency awarded $270 million to four companies working toward building replacements for the space shuttle.

Supporters of the program assert that the competitive nature of the marketplace will help to produce the best ships and launchers at the lowest price, thereby driving down the cost of both manned and unmanned spaceflight. Critics, meanwhile, fear the competition to produce a cheaper spacecraft will put astronauts' lives at risk. This concern over private industry's need for a cost-effective bottom line is not a new one. It recalls something Alan Shepard said after being asked what he had been thinking as he sat in his capsule just before his historic first flight: "The fact that every part of this ship was built by the low bidder."

Private aerospace companies have always been part of the U.S. space program. The Atlas and Titan rockets that first sent Americans into orbit in the 1960s were originally built by private companies that designed launchers for nuclear warheads. It was only later that they were adapted for manned spaceflight. Even the vaunted Saturn V rockets that took astronauts to the moon—which were designed specifically for spaceflight—were contracted out by the space agency. Why, then, are critics wary of Obama's plan to let private industry design and build the next generation of manned spacecraft and launch vehicles?

At the time Shepard made his quip, NASA's rocket scientists and engineers were responsible for coming up with designs and supervising outside contractors. Under Obama's plan, NASA would buy ready-made spacecraft and rockets from companies overseeing their own design, testing, and development phases. NASA would kick the tires, so to speak, but the agency would be the buyer, not the developer.

Some of the companies submitting proposals for the Commercial Crew and Cargo Program are veterans of aerospace engineering. Boeing, for example, is the third-largest aerospace and defense contractor in the world and has worked on numerous projects for NASA. Under the program, it's testing the CST-100 crew capsule, which may be flown atop the Atlas V, Delta IV, and Falcon 9 rockets. (The CST-100 is similar to the Orion capsule originally proposed for Project Constellation. Orion is still under development at Lockheed Martin and is anticipated to make its first manned flight in 2016.) Other aerospace companies like SpaceX, which is owned and directed by PayPal co-founder Elon Musk, are relatively new. For its entry, SpaceX plans to use Falcon 1 and Falcon 9 rockets to launch its Dragon capsule into orbit. Already, the company has had success. In December 2010, SpaceX became the first private aerospace firm to launch, orbit, and recover a spacecraft.

Before forming an opinion on private spaceflight, it's crucial to learn about these companies and the people who run them. The articles in this section examine some of the firms vying to do business with NASA. The author of the first article, "Businesses Take Flight, with Help from NASA," looks into Mark Sirangelo's Sierra Nevada Corporation, as well as SpaceX, Boeing, Orbital Sciences Corporation, and Virgin Galactic, a subsidiary of Richard Branson's Virgin conglomerate that plans to ferry paying tourists to orbit. In "Not Lost in Space: IE Seeks Cost-Effective Means for Suborbital Tourism," the subsequent selection, Michael Hughes writes about Brian Feeney's DreamSpace, which plans to send passengers into space by 2012 and later develop space carriers capable of flying people from New York to Paris in less than an hour.

Next up, in "The Explorer," Max Chafkin profiles SpaceX founder Elon Musk, delving into both his aerospace initiative and his other revolutionary company, Tesla Motors, which produces electric cars. In the following article, "Moon Dreams: The Americans May Still Go to the Moon Before the Chinese," a writer for *The Economist* looks at how private companies might land Americans back on the moon. Finally, in "Commercial Crew and NASA's Tipping Point," Jeff Foust discusses the inevitability of private spacecraft, noting that "commercial crew is increasingly not a question of if it will exist, but how big of a program it will be and how soon it can field systems."

Businesses Take Flight, With Help From NASA*

By Kenneth Chang
The New York Times, January 31, 2011

Sitting in a testing facility at the University of Colorado, the inner shell of the Dream Chaser space plane looks like the fuselage of an old DC-3.

The test structure has been pushed and pulled to see how it holds up to the stresses and strains of spaceflight. With an additional infusion of money from NASA, the company that makes the Dream Chaser, Sierra Nevada Space Systems, hopes to complete the rest of the structure and eventually take astronauts to orbit.

"Our view is if we could stop buying from the Russians, if we could make life cheaper for NASA, and if we could build a few vehicles that do other things in low-Earth orbit that are valuable, isn't that, at the end of the day, a good thing?" said Mark N. Sirangelo, the company's chairman.

The Dream Chaser is one of several new spacecraft that companies are hoping to launch into space with help from the government. Last year, the Obama administration pushed through an ambitious transformation for NASA: canceling the Ares I rocket, which was to be the successor to the current generation of space shuttles, and turning to the commercial sector for astronaut transportation.

So far, most of the attention in this new commercial space race has focused on Boeing, which has five decades of experience building spacecraft, and Space Exploration Technologies Corporation—SpaceX, for short—a brash upstart that gained credibility last year with two launchings of its Falcon 9 rocket.

SpaceX, led by Elon Musk, a founder of PayPal and chief executive of Tesla Motors, already has a NASA contract for delivering cargo to the space station, and says that it can easily add up to seven seats to its Dragon cargo capsule to make it suitable for passengers. Boeing is also designing a capsule, capable of carrying seven passengers, under the corporate-sounding designation of CST-100.

But Boeing and SpaceX are not the only competitors seeking to provide space taxi services, a program that NASA calls commercial crew. Last year, in the first-round financing provided for preliminary development, Sierra Nevada Space Systems won the largest award: $20 million out of a total of $50 million.

In December, another space company, Orbital Sciences Corporation, announced it had submitted a similar bid for a space plane it wants financed during the second round. NASA is to announce the winners by the end of March, and they will divide $200 million.

About half of NASA's $19 billion budget goes toward human spaceflight—the space shuttles, the International Space Station—and $200 million this year is just a small slice.

"If this is indeed the path to do this work, it's probably not what they should be putting into it," said Mr. Sirangelo, who is also chairman emeritus of the Commercial Spaceflight Federation, a trade group. "But on the other hand, it's a lot more than we had before. And it's an acknowledgment there's momentum in the industry and what we're trying to accomplish. So that's good."

After the second round, NASA would like [to] narrow its choices down to two, maybe three, systems to finance.

"We think this is in effect a one-year race to see who gets the furthest," Mr. Sirangelo said, "and at the end of that, presumably the next two years of the authorization bill gets funded, and then you compete for that pot of money."

The blueprint for NASA, passed by Congress last year and signed into law by President Obama, calls for spending on commercial crew to rise to $500 million each year in 2012 and 2013.

Senator Bill Nelson, the Florida Democrat who was one of the primary architects of the blueprint, as the authorization act was called, has said the intent was to provide $6 billion over six years.

But what Congress puts into the budget could be far less.

"They're not getting $6 billion over six years for commercial crew," said a Senate aide who was not authorized to speak for attribution. "That's never going to happen."

The aide estimated commercial crew might receive half that much.

In addition, Congress has not passed the final 2011 budget, and Mr. Obama wants to freeze spending at many federal agencies. Whether the freeze includes NASA will not be known until the president's 2012 budget request is released in two weeks.

While Sierra Nevada has the lowest profile of the companies seeking commercial crew business, it is not new. The parent company, the Sierra Nevada Corporation, is a privately held defense electronics firm founded in 1963, and a few years ago, it bought several space companies and rolled them into the space systems subsidiary.

The space systems subsidiary, located outside Denver, is the largest manufacturer in the United States of small satellites, Mr. Sirangelo said. The satellites, used for communications and other purposes, cannot do everything that large ones can do, but what they can do, they do more cheaply and more efficiently.

The Dream Chaser embodies the same philosophy. "There are some tasks that can be done by smaller, cheaper vehicles that used to be done by very expensive vehicles," Mr. Sirangelo said.

Mr. Sirangelo said the company had invested its own money into the Dream Chaser—indeed, more than the $20 million that NASA has provided. Over the past year, the company has done a test-firing of the engines it plans to use on the Dream Chaser, and it dropped a scale model of the spacecraft from a helicopter to verify the aerodynamics.

But it is a jump from making spacecraft components and small satellites to building a crew-carrying space plane, and where Sierra Nevada lacks in skills and experience, it has brought in other companies and institutions. Its Dream Chaser partners include Draper Laboratory, which has been designing spacecraft guidance systems since Apollo; NASA's Langley Research Center, which did much of the development that the Dream Chaser is based on; Boeing, which has also worked on space planes; and United Launch Alliance, a joint venture of Boeing and Lockheed Martin that builds the Atlas V rocket that the Dream Chaser would ride atop.

Virgin Galactic, the spacecraft division of Sir Richard Branson's Virgin empire, signed on as a strategic partner in December. Among the possible roles that Virgin could play is selling seats on the Dream Chaser. (Virgin signed a similar agreement with Orbital.)

The design of the Dream Chaser also has a long lineage, inspired by a Soviet spacecraft. In 1982, an Australian reconnaissance airplane photographed a Russian trawler pulling something with stubby wings out of the Indian Ocean. It turned out to be a test flight of a space plane called the Bor-4, and the craft captured enough curiosity that engineers at NASA Langley copied it.

NASA called its version the HL-20, and for a while in 1991, it looked to be the low-cost choice for taking astronauts to and from the space station. Then the NASA administrator who liked it, Vice Adm. Richard Truly of the Navy, left, and the man who replaced him, Daniel S. Goldin, thought it was not cheap enough and ended the work.

The Dream Chaser design keeps the exact outer shape from the HL-20—a decision that allows Sierra Nevada to take advantage of years of wind tunnel tests that Langley had performed—while modifying the design within. The biggest change is the addition of two engines, which reduces the number of passengers to seven from 10, but adds maneuverability. To finish developing the Dream Chaser would require less than $1 billion, Mr. Sirangelo said, and it could be ready to fly an orbital test flight in three years.

He imagines that one flight could combine a number of tasks—taking astronauts to the space station and then stopping on the return trip to repair or refuel a satellite. "This vehicle is perfectly designed to do all that," Mr. Sirangelo said.

Officials at Orbital Sciences—a company in Dulles, Va., that builds and launches rockets and satellites for everything from television broadcasts to scientific research—say they are excited by the possibilities of commercial crew, but they are more cautious. Orbital, founded in 1982, was a survivor from the last boom-and-bust in commercial space.

Its space plane design is a refinement of the HL-20. Following in the pattern of tapping Greek mythology for the names of its spacecraft, Orbital calls its plane Prometheus. Orbital says development of Prometheus would cost $3.5 billion to $4 billion, which would include the cost of upgrading the Atlas V rocket and two test flights.

With enough financial support, David W. Thompson, chief executive of Orbital, is sure that his company can build and operate Prometheus. But he is less sure that his industry is at a tipping point for spaceflight to become much more common, driving down prices and opening up space to new businesses.

"I think it depends on what the demand curve really is," Mr. Thompson said. "I would say I'm highly skeptical."

Not Lost in Space*

IE Seeks Cost-Effective Means for Suborbital Tourism

By Michael Hughes
Industrial Engineer, July 2010

Twenty thousand bucks buys you a car these days. In a few years, it could purchase several minutes in space.

That's all because industrial engineer Brian Feeney read about the X Prize in a magazine in 1996. The $10 million award was for the first nongovernment organization to launch a reusable manned craft into space twice within two weeks.

"I said that's it," Feeney recalled. "I turned my battleship left and basically started working on those designs that day."

Since then, Feeney has used his decades of experience in product development, production, manufacturing and design to pursue manned space flight in a cost-effective manner. Although his da Vinci Project team didn't win the X Prize, it set the foundation for the DreamSpace Group. The traditional industrial engineering zeal for cutting costs while maintaining safety and quality allows the Toronto-based concern to compete with more well-funded would-be space travelers Virgin Galactic, which already is test-flying. Feeney said Virgin's investment is in the $200 million to $300 million range, while DreamSpace relies on about $20 million to $30 million.

Feeney, the chairman, president and CEO of DreamSpace, calls Virgin a financial juggernaut and credible competitor plowing the road for the entire space tourism industry.

"Our means of competing with them obviously is something that is more boutique, but I think a better experience," Feeney said. "Just think of yourself being in a large telephone booth 6 feet wide or maybe a bus stop or something like that. And you've got a view on three sides, so instead of looking through a portal you've got this spectacular view of space, and you're really going to feel like you're floating in space."

DreamSpace plans to charge $89,000 to fly one passenger about 140 kilometers above sea level. Single-passenger tours will last about one hour and 15 minutes, spending about seven minutes in space. For $49,900 apiece, two passengers could fly about 110 to 120 km above sea level.

"Initially, we're saying over 100 km, but we expect it to be in the 110 to 120 range," Feeney said. "And that will depend in part on if you've got a couple of linebackers from the Pittsburgh Steelers in there. If it's a couple of them, I might say you've got to go alone."

The company wants to carry its first passenger up in mid- to late 2012, about a year behind Virgin Galactic. Five to 10 years later, Feeney hopes to use an enhanced engine to add a third passenger, dropping the price to $19,900 per person. The size of the XF1-B model's cabin—6 feet wide, 5.5 feet high and 10 feet long—easily can accommodate three people, he said.

"It's not about technology," Feeney said. "It's about our ability to produce these things in a certain volume—six to 12 spacecraft a year is our goal—and have the economies of scale we believe is there. But until we get going we can't absolutely guarantee that."

Like Henry Ford selling Model Ts to the general public, DreamSpace plans to sell XF1-Bs to those who want their own space plane. Eventually, DreamSpace could separate into two companies, one for manufacturing and one for space tourism.

Meanwhile, the XF1-A prototype, which will carry a pilot only, should roll out later this year. Current modifications include larger cycle engines and adding more cabin room so passengers can float. Unlike Virgin Galactic, which will use a mother ship to launch the spaceship at altitude, XF1s will fly from their own runway. The changes should help develop a separate aircraft to compete in the supersonic business jet category without reaching suborbital heights. The long-term goal is point-to-point suborbital travel, although that could take several generations of aircraft.

"While this will take some time to develop, we have a midrange goal of demonstrating New York to Paris in 59 minutes, which requires an average speed of just under Mach 6," Feeney said.

By approaching space travel in a different way, DreamSpace is, from the start, designing low-cost flights with a good capital return, he said.

The Soviet Union and NASA launched people with the help of intercontinental ballistic missiles, just like the soon-to-be-dormant NASA shuttle program. Japan and China are taking the same path, which chucks millions of dollars worth of hardware into the ocean each launch.

"Every time you build one of these things it's an experimental flight," Feeney said. "Even if you've done 10 of them you can't say for certain that everything is now hunky dory because you are throwing it away every time. You do not have the ability to take parts to the lab and say, 'Hey do an X-ray analysis.'"

Feeney didn't want to redo what governments have spent billions on. And he knew DreamSpace needed better than the 98 percent success rates of government-run space vehicles.

"I couldn't put someone on board, including myself, on something that had a 98 percent success rate," he said. "You've got to be one in 10,000, one in 100,000, but you can't get there in terms of your statistical analysis until you've done hundreds of flights. And even then you'll have to stretch your numbers, but you'll have some evidence and some idea of what can go wrong and what systems might be failing along the way."

To improve safety, Feeney's team is designing the XF1s with the idea that failure will happen.

"As sure as God made 747s, you're going to have problems," he said. "Hardware will fail, from a minor to a major problem. Our whole approach is 'Don't lose the people.' "

The spacecraft has passive static stability for re-entry. If the engines don't fire at 50,000 feet, the plane comes back and lands. The entire vehicle has a parachute system. The entire cockpit can detach and parachute down, "almost like a giant bathtub," at any time during the flight profile. And while the last thing they want to do is have people jump out, there's a system for that, too.

"You can't modify aircraft to do this," Feeney said. "But we took a white sheet of paper, which is the neat way of doing it, and we designed all that from scratch."

The smaller XF1-A will cost-effectively develop and improve these technologies and flight controls.

"When you scale it up by 100 percent, your costs go up dramatically for a variety of reasons," Feeney said. "So we're going to fly at low levels, get it to operate as an aircraft first and foremost to fly in stable flight and then develop its flight envelope all the way up to space."

Although the prototype will fly and be tested hundreds of times, it should take only 30 to 50 flights to get to space. That gives DreamSpace the credibility to bring in more capital and fully develop the commercial XF1-B. The A model will continue as a test bed, evolving internal systems like life support so the team can make small changes before applying them to something larger and more costly.

Feeney also keeps costs low by using a mix of expensive materials (exotic carbon fibers) when necessary and off-the-shelf items (commercial valves) when possible. One brand of valves used from the Apollo missions onward cost $10,000 to $15,000 apiece. DreamSpace modifies $150 valves to meet the spacecraft's criteria.

"We don't need the same fast millisecond blinking open unit and shut time frame because we didn't design the system to have that need," Feeney said. "So you're looking at an order of magnitude to dropping costs, and so when things need to be maintained, they can be maintained cost effectively."

Feeney said DreamSpace can produce a rocket engine for about $10,000, even though the large dynospace companies charge governments millions for the same thing. But competition, improvements in machining and good industrial design allow DreamSpace to design out a lot of the manufacturing costs.

Feeney said DreamSpace is not about performance, but about reliability. It's not about the lightest weight, but "the lightest weight that we can achieve so long as we don't go over the barrier that says thou shalt not cost more than this," he said. "And if you have to make the vehicle a little bit bigger, a little bit heavier because you're carrying a little more weight of systems, well, then so what."

The testing, retesting and the drive to learn the manufacturing process is helping develop commercial technologies and apply the best industrial design practices to come up with a reasonable structure for repeated suborbital trips. Feeney said the point of suborbital space tourism is to let the public at-large cost effectively get out of the cradle and have a view of about 1,500 km in either direction. From a perch 140 km above Toronto, space tourists easily could see Boston, New York, Philadelphia, Washington and the Great Lakes.

"Our approach is to not rush into orbit," Feeney said. "My goal is to get to orbit and stay there, and get to the moon and Mars, and all the lofty goals that any kid has ever had. Or at least in the case of Mars, I might be laying the foundations for people to cost effectively do it in the many decades to come."

The Explorer*

By Max Chafkin
Inc., October 2010

There are many things of which Elon Musk is certain. Musk knows, for instance, how to get a liquid-fueled rocket into orbit and how to run a sports car with 7,000 tiny batteries. He knows how to manage 1,700 employees spread between two wildly ambitious companies in two different cities: the Los Angeles aerospace company SpaceX and the Bay Area electric-car company Tesla Motors. He knows how to get rich.

And then there are things Elon Musk doesn't know but simply *believes*. He believes that Northrop Grumman couldn't build an inexpensive rocket even if someone literally handed it plans to do so. He believes that the Chevy Volt will be a disappointment when it goes on sale later this year. And he believes that in roughly 20 years, he will step out of a space capsule and become one of the first humans—part of a new generation of explorers, without precedent since the days of Columbus and Magellan—to establish a human colony on Mars.

The last, of course, will require hard work. Musk will have to create and then launch a rocket capable of safely transporting humans beyond Earth's orbit—all while avoiding the machinations of his competitors, the lingering effects of the global recession, and the ill will of members of Congress and the public who are annoyingly hostile to the idea of having private companies set up colonies on other planets.

Yes, getting to Mars will be a challenge, even for someone who knows as much as Elon Musk. So he isn't worrying much about getting back. His will most likely be a one-way mission: a glorious and romantic and—let's be honest—insane attempt to take civilization beyond this planet.

Whatever one thinks of Musk's ideas about multi-planetary life, he is a singular character in American business. Musk has already helped inspire a Hollywood action hero—Tony Stark, portrayed by Robert Downey Jr. in the *Iron Man* movies—and has hundreds of pages' worth of press clippings. He tends to remind people of

Apple co-founder Steve Jobs. The comparison has merit: both men are committed micromanagers, and both have an instinct for the theater of business. But whereas Jobs is very good at working within constraints—designing the perfect cell phone or the best computer operating system—Musk likes to aim his energies far beyond the normal limits of common sense. "Lots of successful entrepreneurs want to change the world," says Steve Jurvetson, a longtime venture capitalist who has invested in Tesla and SpaceX. "For Elon, that's too narrow." The solar system is the thing.

These grand ambitions date at least to college, when Musk had an epiphany about what to do with the rest of his life. "I decided that there were three areas that were most going to affect the future of humanity: the Internet, sustainable energy, and extending life beyond Earth," Musk tells me. "I'm sure you've heard this story before."

I had heard this story before. In fact, in some small way, I helped create it. Three years ago, I wrote an article for *Inc.* about Musk, who had already started and sold two successful Internet companies. When he was just 23, Musk co-founded Zip2, an early Web software company that Compaq bought for $300 million, and at 27, he helped start PayPal, the Internet payments company for which eBay paid $1.5 billion. These accomplishments made him absurdly wealthy, but when I met him in 2007, he hardly seemed qualified to colonize another planet or cure humanity's addiction to oil. And yet for some reason I—along with many others who heard his story—believed he could do it.

Why that was, I'm not sure. Maybe it was his background, which reads like a heroic myth of entrepreneurship. As a boy growing up with divorced parents in apartheid South Africa, Musk dreamed of escaping to America. When he was 17, he left home and enrolled in college in Canada. He eventually graduated from the University of Pennsylvania. In 1995, he was accepted into a Ph.D. program in materials science and applied physics at Stanford, only to decide, two days after he had arrived on campus, to drop out and start a company.

Or maybe it was the oversize scale on which Musk seemed to live his life: not a family, but five boys, all under the age of 5. Not just one company, but three: SpaceX, Tesla, and SolarCity, a solar panel company that Musk dreamed up, funded, and then turned over to two of his cousins, Lyndon and Peter Rive.

Or maybe it was his physical presence: the imposing frame—Musk stands over 6 feet tall, is thick through the middle, and carries himself like a rugby player—the haughty South African accent, and the striking, if not exactly handsome, face. Or maybe it was the fact that nearly everybody who had ever worked with him, even people who clearly despised him, seemed to think he was a genius.

By late 2007, thanks to Musk's guidance and cash, Tesla Motors had built a prototype for an electric car that, it was said, could beat a Ferrari off the line. His spaceship had yet to reach orbit—its ultimate goal—but there had been two launches and a dozen signed contracts, including a $278 million deal with NASA. [Musk's Dragon space capsule reached orbit on December 8, 2010.] SpaceX's was the first privately funded rocket that seemed to have a legitimate chance at competing with those developed by the government and operated by the big aerospace companies, and it promised to reduce the cost of a launch dramatically. (SpaceX charges $50 million to launch a satellite, less than half the going rate.) Finally, SolarCity had quickly become the dominant solar panel installation company in California, with $23 million in revenue and nearly 200 employees.

On the strength of these accomplishments, *Inc.* named Musk the 2007 Entrepreneur of the Year. But to me, even this accolade felt like an understatement. Musk seemed like a superhero—or maybe an alien. He told me that what he was doing would be one of the biggest events in the history of humanity, at least on a par with the moment when our forebears flippered their way out of the ocean and began walking on dry land. He said this seriously.

I returned to California this summer to see how the second act of Elon Musk's story was playing out. It was a Tuesday night at the Tesla Motors headquarters in Palo Alto, and Musk was in fine form, taking meeting after meeting about various aspects of the design of Tesla's forthcoming sedan, the Model S, and cracking jokes. He had just returned from his first vacation in eight years—four days in French Polynesia and another four in a rental in Brazil after visiting a cousin. "It was the worst place I've stayed since I was a teenager," he says, shaking his head and letting out a laugh. "It stank, there were stains on the sheet, and the bedbugs did bite. It was awesome. We had a good time."

The vacation was a celebration of sorts. Three weeks before, I had watched Musk ring the Nasdaq bell in Times Square. Because Nasdaq companies are traded by computers rather than by guys in numbered blazers, there is no actual stock exchange—just a cramped television studio in New York City—but Musk hadn't let that fact get in the way of his spectacle. He showed up wearing a purple blazer and a checkered shirt, flanked by his beautiful 25-year-old fiancée, the actress Talulah Riley, and his now-6-year-old twins, Griffin and Xavier.

Musk noted, in his opening remarks, that this was the first IPO for a car company since Ford went public in 1956. He rang the bell, pumped his fist, and then took his entourage outside to pose for photos with Tesla cars. "We've confounded the critics at every turn," he bragged to CNBC, as its cameras panned to a bright red prototype of the company's Model S, a luxury sedan that will sell for roughly $50,000. "At a certain point, people have to get tired of being wrong."

The spectacle worked: Despite skepticism from market pundits—case in point, *Mad Money*'s Jim Cramer, who, just before the IPO, crowed, "You don't want to own this stock! You don't want to lease it! You shouldn't even rent the darn thing!"—Tesla's stock settled $3 above the offering price of $17 a share, giving the company a market capitalization of roughly $2 billion. (Musk sold 5 percent of his stock in the IPO but remains the company's largest shareholder, with 30 percent of its stock.)

Musk spends two to three days a week in Palo Alto, flying in late on Tuesday mornings. He works pretty much nonstop until he flies back to Los Angeles, where he lives. When he must pause to eat, he does so with amazing efficiency. I twice saw him consume an entire meal—chicken, a vegetable, bread, and a Diet Coke or two—in under five minutes, all while holding forth on topics such as how best to fix a rocket vibration problem or the ridiculousness of sales as a business function. "In the early days, when Elon would have lunch meetings, I used to have to tell people that they shouldn't worry if he'd already finished before they even sat down," says Mary Beth Brown, his longtime assistant. "He just doesn't realize how fast he's moving."

Musk needs those precious minutes. Despite the growth of Tesla and SpaceX, which have tripled their staffs since 2007, Musk hasn't really changed his management style. He still vets almost all new employees—though lower-level hires are now allowed to answer his interview questions by e-mail—and he remains the lead product designer for both the rocket and the electric car. "A normal workaholic is sober compared to Elon," says Lyndon Rive, the CEO of SolarCity, for which Musk serves as chairman and in which he holds a 25 percent equity stake.

Musk's favorite activity is to lead technical meetings at Tesla and SpaceX. There are dozens of these gatherings each week on topics such as the car's battery pack or the rocket's navigational software. Typically, Musk sits at his desk with a dozen or more young engineers spread out on chairs, the windowsill, and the floor. Managers attend these meetings, too, but they don't do much talking; Musk prefers to get his information from the kids doing the actual work. During one of our five-minute lunches, I ask him how many people report to him. "It's not a zillion," he says, then recites a few names and guesses that it's about 20. "It's probably more than I think it is."

But reporting structures don't matter to Musk, who has trouble staying out of any detail, no matter how small or seemingly trivial. The new Tesla headquarters, a sprawling 400,000-square-foot campus that used to be a Hewlett-Packard laboratory, was still under construction when I visited, and Musk, not one to miss a bunch of potentially interesting decisions, immediately inserted himself into the process. He cast the deciding vote in a debate over how much red paint to use in the warehouse, he closely scrutinized a plan for the staircase railings, and he engaged in a 10-minute colloquy about the proper material for restroom countertops. (The director of facilities recommended tiles; Musk ruled in favor of a granite slab.)

I had a hard time keeping a straight face while the self-taught rocket scientist got worked up about bathroom furnishings. But later that week, when I used the SpaceX men's room, I noticed a design touch that seemed positively Muskian. When I asked him about it, Musk informed me that he had indeed personally selected the toilets at SpaceX. His favorites: a urinal that incorporates a psychedelic strobe light and another that employs, instead of a porcelain receptacle, a large steel bucket. "I was looking for creative urinals," he explains. "It's a nice pot."

The allusion to having a pot to piss in is apt. Two years ago, Musk and his companies were perilously close to bankruptcy, and Musk was perilously close to falling apart. "Those were dark days," he says of the summer of 2008. "I think I still have some emotional scar tissue, just thinking about that time stresses me out."

The trouble began in 2007, just as production for the Tesla Roadster was set to begin. That was when Musk discovered that the Roadster, which was supposed to sell for $92,000, was actually costing the company $140,000 in raw materials alone. Musk blamed the failure on lax accounting practices—and on intentional obfuscation by co-founder and CEO Martin Eberhard. He fired Eberhard and installed a new CEO, who began renegotiating supplier contracts, slashing costs, and raising prices. Over the next year, Tesla would fire roughly 30 percent of its staff and close the Detroit office that had been developing the Model S sedan. Eberhard did not leave quietly. On his blog, Teslafounders.com, he portrayed Musk and his cohorts as coldhearted, shortsighted, and mean, calling the layoffs a "stealth bloodbath." (The pair eventually resolved their dispute in mediation.)

In addition to the public spat with Eberhard, Musk found himself embroiled in a lawsuit with Henrik Fisker, a former Tesla designer who had started a competing company. Meanwhile, Musk's marriage was falling apart, and his wife, Justine, being a professional writer, was, naturally, blogging about the divorce. "No one who knows my ex-husband would accuse him of being weak-willed," she wrote that October. "The same qualities that helped bring about his extraordinary success dictate that the life you lead with him is his life . . . and that there is no middle ground (not least because he has no time to find it)."

The result of all this was a barrage of terrible press. A long article in *Fortune* questioned whether Tesla would ever deliver production cars. A series of posts on the blog Valleywag suggested that Musk was intentionally steering Tesla into bankruptcy, that he had engaged in an extramarital affair, and that he was a habitual liar. Musk did his best to strike back with denials and rejoinders, but the battling took its toll. "I'd never seen him so sad," says Maye Musk, Elon's mother. "Everything was collapsing around him."

It got worse. In August 2008, SpaceX suffered its third consecutive launch failure, losing a rocket and two satellites—and the remains of James Doohan, the actor who played Scotty in the original *Star Trek* TV series. Doohan had paid to have his ashes shot into orbit. Instead, they ended up in the South Pacific, with what was left of the rocket and the satellites.

Musk was devastated. SpaceX's goal had been to make rockets cheaper to launch and more reliable. Instead, the company had lost every rocket it had launched and had spent nearly all of the $100 million that Musk had used to found the company.

Then, just as Musk was trying to raise additional funds so Tesla could begin building the Model S, the credit markets collapsed, and the auto industry seized up. Tesla had just four months of operating capital left in the bank at a time when no investor wanted to sink money into an unprofitable car company. Even SolarCity was foundering. Morgan Stanley, which had been financing no-money-down solar panel leases, pulled out of the program, temporarily cutting the company's sales in half. "It looked like all three might go under," says Musk.

I had never seen Musk show any kind of emotional vulnerability before he told this story. He looked down and then confessed that he would often wake up and discover that he had been sobbing in his sleep. To Musk, a man who had steered his own destiny since his early teens, the loss of self-control was terrifying. "I'd wake up and just be like, What the f***," Musk says. "I think that it's like"—he pauses, suddenly seeming far younger than his 39 years, and then tries to explain—"When you're asleep, there's just much less emotional control."

After SpaceX's third launch failure, Musk announced that the company was raising outside capital for the first time—$20 million from the Founders Fund, a venture capital fund run by his former PayPal cofounder Peter Thiel. The investment did two things: It showed that somebody other than Elon Musk believed he might succeed, and it gave him enough money for at least two more launches.

Rocket companies normally spend months after a launch failure carefully reviewing what happened before they make their findings public. But just four days after the crash, Musk took to the SpaceX blog and wrote that he was "certain as to the origin of this problem." He planned to have a new rocket on the launching pad within a month. To industry veterans, the swiftness of the response was impressive—and also a little frightening. "A lot of people were seriously questioning whether SpaceX could put everything together and have a successful launch," says Jeff Foust, an aerospace analyst with the Bethesda, Maryland, consulting firm Futron and the editor of *The Space Review*, an online journal that covers the industry. "If Elon had failed again, he would have been roundly criticized for not taking a more methodical approach. It was a bit of a bet on his part."

But SpaceX didn't fail. On September 28, 2008, Elon Musk's *Falcon I* rocket became the first privately funded spaceship to reach orbit from the ground. A grainy webcast shot from a camera mounted on the rocket showed the engine burning red above the blue Earth, while loud cheers from hundreds of employees echoed over the feed. In December, NASA announced that it had purchased 12 flights on Musk's new, larger rocket, the *Falcon 9*, to resupply cargo to the International Space Station; the contract was worth roughly $1.6 billion over seven years.

Musk made a similarly risky gamble to shore up Tesla. At a board meeting in October, with the company's future in doubt, Musk informed the board that he was raising a $40 million round from existing shareholders—even if it meant that he would be the only shareholder putting in money. He eventually invested about $20 million in cash. "That was it," he says. "I was all in."

Musk's investors were floored. "I don't think I've ever seen an entrepreneur with so much resolve," says Steve Jurvetson. "That was a heroic act. It was a risky act. It saved the company." The gesture helped persuade Jurvetson and other Tesla shareholders to dig a little deeper. Kimbal Musk, Elon's brother and a board member at Tesla and SpaceX who contributed to the round, told me that it never occurred to him to advise his brother not to bet the remainder of his wealth on Tesla. "There was no question he was going to do it," says Kimbal. "Elon's psyche is so tied up in the idea of changing the world. His attitude was, So what if I don't have any money left?"

Musk's official titles at Tesla had been product architect and chairman, but as the largest shareholder, he had long been a de facto CEO. Now he took the title formally and announced that he would personally refund buyers' deposits if Tesla failed to produce a car. He also began aggressively campaigning to tap several hundred million dollars' worth of government-guaranteed loans that had been approved as part of the $7.5 billion Advanced Technology Vehicles Manufacturing Loan Program, a green-auto initiative that President George W. Bush signed into law just before leaving office.

The decision to apply for government funding was controversial, especially in the light of the financial crisis and the auto industry bailout. *New York Times* columnist Randall Stross wrote a scathing column in November in which he questioned whether it was appropriate for taxpayers to help out a sports car manufacturer and proposed the loan program be renamed the Bailout of Very, Very High-Net-Worth Individuals Who Invested in Tesla Motors Act. Musk had always been an enthusiastic patron of the press—he accepts most interview requests, rarely seems to censor himself, and does not use a publicist—but he took an especially aggressive tack in defending Tesla during the financial crisis. "Randy Stross is a huge d***he bag *and* an idiot," he said in a video interview with Yahoo's finance website. (Musk stands by the assertion. In fact, when I asked him about it, he spent a few minutes parsing the difference between the two slurs and then he added another: "Renowned d***head." I didn't ask for an explanation, but I'm sure he had one ready.)

Musk defended Tesla's application for the loan guarantee by pointing out that government dollars were already going to gas-guzzling carmakers. How, he asked, could anyone begrudge a tiny electric-car company a few million dollars when Washington was giving billions to General Motors and Chrysler? The argument worked. Tesla won approval for a $465 million loan, the chief source of financing for the Model S. (The big companies made out just fine. As part of the alternative-fuel program, Ford received $5.9 billion, and Nissan got $1.6 billion.)

The confrontational pose has since become Musk's default. During my visits, he made cracks about the astronaut (and vocal SpaceX opponent) Neil Armstrong, Audi, GM, BMW, Democrats, Republicans, the U.S. Senate, Lockheed Martin—as well as a smattering of lesser villains that included enterprise software companies (useless, overpriced) and Santa Monica restaurateurs (purveyors of low-quality fare). These barbs can be quite funny—and they make for excellent copy—but Musk's tendency to publicly humiliate anyone who stands in his way can also be coercive and a little creepy.

On the other hand, Musk has little choice but to defend himself. He points out that his rivals in the auto and aerospace industries are among the most entrenched in the world. "The lobbying power they have is gigantic," he says. "They have literally buildings full of lobbyists in D.C. We have one guy. If this was something about who has the most lobbying power, we would be screwed." Musk has received plenty of government money—and the Obama administration has embraced Tesla and SpaceX—but suggestions by commentators that this is the result of some kind of quid pro quo arrangement seem off base. During the 2008 election, Musk, a registered Independent, contributed more money to Republicans than Democrats. He gave $2,300 to the Obama campaign.

Still, his days as an outsider are probably numbered. Tesla is on track to book about $100 million in revenue this year, on sales of the Roadster and battery packs for Daimler's electric Smart car and the Mercedes A-Class—as well as Toyota's electric Rav4. (Both automakers now have substantial investments in Tesla; the battery-pack deals are joint ventures.) In 2012, the Model S is set to start rolling off production lines in a Bay Area factory that once made Chevy Novas and Geo Prizms for GM.

SpaceX, meanwhile, has 40 launches on its manifest—including the NASA deals, a $500 million contract with the satellite giant Iridium, and contracts to provide launch services to the space agencies of Taiwan and Argentina. "For many years, we've seen proposals for lots of new launch vehicles, but most of them never materialize," says Foust, the aerospace analyst. "For SpaceX to have launched the *Falcon 9* successfully is a major accomplishment. Commercial space has moved past the PowerPoint stage."

Between the electric cars and rocket launches, Musk clearly has a lot on his plate. But that hasn't stopped him from plotting another trio of ambitious business proposals.

"The simplest," Musk tells me, "is to use aerospace engineering to double-decker the freeways." He plans to start a company that will create prefabricated metal risers out of space-grade aluminum. Then, he wants to drop those risers onto Los Angeles's famously congested 405, doubling the capacity of the freeway without stopping traffic. "It's a no-brainer—easily done," he says. "They would also look quite pretty."

His other two ideas are more speculative, but not entirely so. Musk hopes to one day turn his attention to solving the problem of commercial nuclear fusion—that is, generating nuclear power without nuclear waste—and to design and manufacture a zero-emission airplane. "The idea would be to go several generations beyond what's currently available," he says. "It would be an electric, supersonic vertical-takeoff plane—much more convenient. I know that would work, too."

It sounds pretty wild, but so did Musk's ideas when, in 2002, he looked into shooting a greenhouse into space and landing it on Mars. But after making three trips to Russia in an attempt to buy a cut-rate launch on a decommissioned intercontinental ballistic missile, Musk learned that even the least expensive rocket launch would cost $20 million or more. "The cost was prohibitive," he says. "It's 100 times worse than it needs to be, and it hasn't been improving. The U.S. had almost no rocket infrastructure in 1960. Nine years later, we were on the moon. And so you'd think, Sure, we can go to Mars by now. No way. We can't even get back to the moon."

SpaceX is working on a version of its rocket that will use two extra boosters to carry as much as 35 tons into orbit. Musk says the parts necessary to build a Mars spaceship could be launched on these more powerful rockets and then assembled in space, the same way the International Space Station was built. Or SpaceX could build a much larger craft—"A big freaking rocket," Musk says—to go straight to the Red Planet from our own.

None of this, of course, is paid for yet, and there is no clear market for Mars rockets. "We're taking a walk-before-you-run attitude," says Lori Garver, the deputy NASA administrator and the former space adviser to the Obama campaign. "We're hoping there will be some near-term successes to prove out our plan for commercial space." To date, *Falcon 9* has flown only once, and there are two test flights remaining before it can begin flying to the space station. If it suffers more failures, any talk of Mars would seem to be off the table.

In that case, it's hard to imagine Musk failing gracefully. The psychic blow would be too great—or he would destroy himself financially trying to salvage his dreams.

Musk hopes to eventually devote the bulk of his energy to SpaceX. Tesla's government loans require him to hold on to most of his stock until the loans are repaid, and an agreement with Daimler mandates that he stay on as CEO at least until the completion of the Model S in 2012. "I'm committed to Tesla, but nobody should be CEO forever," he says. "I think I've burnt out some mental circuits over the past couple of years."

Musk's chief concern is that by the time humanity is ready to get to Mars, he will be too old or frail to survive the journey. "I don't want to be doddering around up there, needing a quadruple bypass or something," he says, adding that, ideally, he will be on Mars by 2030, and by 2040 at the very latest.

But even that sounds ambitious. I ask him if he can really, really get it all done in his lifetime.

"Probably," he says. "Hopefully."

Then he pauses for a few seconds to consider the enormity of that question. He will be 69 in 2040. "Thirty years," he says. "That's me since I was 9. I think I can get a lot accomplished in that period of time."

Moon Dreams*

The Americans May Still Go to the Moon Before the Chinese

The Economist, February 18, 2010

When America's space agency, NASA, announced its spending plans in February, some people worried that its cancellation of the Constellation moon programme had ended any hopes of Americans returning to the Earth's rocky satellite. The next footprints on the lunar regolith were therefore thought likely to be Chinese. Now, though, the private sector is arguing that the new spending plan actually makes it more likely America will return to the moon.

The new plan encourages firms to compete to provide transport to low Earth orbit (LEO). The budget proposes $6 billion over five years to spur the development of commercial crew and cargo services to the international space station. This money will be spent on "man-rating" existing rockets, such as Boeing's Atlas V, and on developing new spacecraft that could be launched on many different rockets. The point of all this activity is to create healthy private-sector competition for transport to the space station—and in doing so to drive down the cost of getting into space.

Eric Anderson, the boss of a space-travel company called Space Adventures, is optimistic about the changes. They will, he says, build "railroads into space". Space Adventures has already sent seven people to the space station, using Russian rockets. It would certainly benefit from a new generation of cheap launchers.

Another potential beneficiary—and advocate of private-sector transport—is Robert Bigelow, a wealthy entrepreneur who founded a hotel chain called Budget Suites of America. Mr Bigelow has so far spent $180m of his own money on space development—probably more than any other individual in history. He has been developing so-called expandable space habitats, a technology he bought from NASA a number of years ago.

These habitats, which are folded up for launch and then inflated in space, were designed as interplanetary vehicles for a trip to Mars, but they are also likely to be useful general-purpose accommodation. The company already has two scaled-down versions in orbit.

Mr Bigelow is preparing to build a space station that will offer cheap access to space to other governments—something he believes will generate a lot of interest. The current plan is to launch the first full-scale habitat (called Sundancer) in 2014. Further modules will be added to this over the course of a year, and the result will be a space station with more usable volume than the existing international one. Mr Bigelow's price is just under $23m per astronaut. That is about half what Russia charges for a trip to the international station, a price that is likely to go up after the space shuttle retires later this year. He says he will be able to offer this price by bulk-buying launches on newly man-rated rockets. Since most of the cost of space travel is the launch, the price might come down even more if the private sector can lower the costs of getting into orbit.

The ultimate aim of all his investment, Mr Bigelow says, is to get to the moon. LEO is merely his proving ground. He says that if the technology does work in orbit, the habitats will be ideal for building bases on the moon. To go there, however, he will have to prove that the expandable habitat does indeed work, and also generate substantial returns on his investment in LEO, to provide the necessary cash.

If all goes well, the next target will be L1, the point 85% of the way to the moon where the gravitational pulls of moon and Earth balance. "It's a terrific dumping off point," he says. "We could transport a completed lunar base [to L1] and put it down on the lunar surface intact."

There are others with lunar ambitions, too. Some 20 teams are competing for the Google Lunar X Prize, a purse of $30m that will be given to the first private mission which lands a robot on the moon, travels across the surface and sends pictures back to Earth. Space Adventures, meanwhile, is in discussions with almost a dozen potential clients about a circumlunar mission, costing $100m a head.

The original Apollo project was mainly a race to prove the superiority of American capitalism over Soviet communism. Capitalism won—but at the cost of creating, in NASA, one of the largest bureaucracies in American history. If the United States is to return to the moon, it needs to do so in a way that is demonstrably superior to the first trip—for example, being led by business rather than government. Engaging in another government-driven spending battle, this time with the Chinese, will do nothing more than show that America has missed the point.

Commercial Crew and NASA's Tipping Point*

By Jeff Foust
The Space Review, February 14, 2011

Later today the White House will release its budget proposal for fiscal year 2012, including its plans for NASA. That release is likely to be far less controversial than the release of its 2011 budget proposal just over a year ago, in which the administration proposed cancelling all of Constellation and replacing it with technology development efforts and an initiative to develop commercial crew transportation systems. That generated a long, often vociferous debate (at least by space policy standards) about the future direction of the agency, culminating with the passage of the NASA Authorization Act that put into law much, but not all, of what the administration was seeking.

This time around no one is expecting similar radical changes for NASA. Instead, the 2012 budget proposal offers the agency an opportunity to further cement the foundation for those new programs, and grapple with the many challenges associated with them, from concerns about the ability of NASA to build a heavy-lift vehicle (see "Can NASA develop a heavy-lift rocket?", *The Space Review*, January 17, 2011) to cost overruns on high-profile programs like the James Webb Space Telescope, to broader challenges of carrying out programs on a budget that will be, at best, frozen at 2011 levels. (Further complicating that last point is that NASA, like the rest of the federal government, still doesn't have a final 2011 budget, four and a half months after the fiscal year began.)

Another high profile program likely to be scrutinized in the budget proposal is the agency's commercial crew development efforts, a topic that was one of the flash points of last year's budget and policy debate. This time around, though, the debate may very well be different. Last year the question was whether there should be a commercial crew development program at all, one that was answered by its inclusion in the authorization act, although at lower funding levels than the administration's request. Now, though, developments both in government and in the private sector suggest commercial crew is increasingly not a question of if it will exist, but how big of a program it will be and how soon it can field systems.

COMMERCIAL READINESS

One of the key issues of last year's debate was the readiness of commercial providers to develop crew transportation systems, something that previously had been done only as government programs with very different contracting relationships with aerospace firms. Those worries were at least partially ameliorated with the successful demonstration flight in December of SpaceX's Dragon spacecraft, developed to carry cargo but intended to eventually carry crew as well (see "2010: the year commercial human spaceflight made contact", *The Space Review*, December 13, 2010).

"The December 8th flight of the Falcon 9/Dragon, for us, demonstrated that the United States commercial sector is prepared to meet the needs of NASA to carry crew to orbit," said Tim Hughes, vice president and chief counsel of SpaceX, during a panel session at the 14th Annual FAA Commercial Space Transportation Conference in Washington last week. Such a system could carry astronauts, he said, at less than half the cost of what Russia is currently charging for seats on Soyuz spacecraft.

SpaceX has submitted a proposal for NASA's second round of Commercial Crew Development awards (known as CCDev-2) to work on what Hughes called the "long pole in the tent" for a commercial crew capability, a launch escape system. If funded, he said that "ideally" the company could have an operational crew capability within three years.

SpaceX is hardly the only company showing an interest in commercial crew development. Several other companies have been developing concepts for spacecraft and/or launch vehicles, either under the first round of CCDev awards made last year or in CCDev-2 proposals now being evaluated by NASA. That interest was evidenced by the standing room only crowd that gathered to attend the panel, which also featured representatives of Boeing, Orbital Sciences Corporation, Sierra Nevada Corporation, and United Launch Alliance (ULA). "I think a few years ago we would have been talking to ourselves," said Mark Sirangelo, chairman of Sierra Nevada Space Systems.

Boeing, Sierra Nevada, and ULA all received first-round CCDev awards last year, and have either completed or are wrapping up work associated with those awards. That work has ranged from completing a system definition review by Boeing of its CST-100 capsule, along with associated subsystem development, to drop tests by Sierra Nevada of a subscale model of its Dream Chaser vehicle, to ULA's work simulating launch failures to test an emergency detection system critical to human-rating its launch vehicles. (ULA also simulated the ascent profile of its Atlas 5 402 rocket using the centrifuge at the NASTAR Center near Philadelphia, with former astronaut Jeff Ashby testing his ability to operate a notional vehicle at accelerations of up to 4 g.)

Orbital, currently developing the Cygnus cargo vehicle for servicing the ISS, jumped into the commercial crew sector in December when it announced it had submitted a CCDev-2 proposal for a lifting body vehicle subsequently named Prometheus. Frank Culbertson, senior vice president at Orbital, said Prometheus is based on design work the company did on NASA's Orbital Space Plane program in the early 2000s. "We think it's going to be a very exciting vehicle for NASA to evaluate and eventually procure," he said.

Perhaps the biggest recent sign of commercial interest in crew transportation development, though, took place last week when ATK announced it had submitted its own CCDev-2 proposal for a new launch vehicle it called Liberty. The vehicle, strikingly similar to the now-cancelled Ares 1, uses the same five-segment solid rocket booster for its first stage. However, the Boeing-built Ares 1 second stage, powered by a J-2X engine, has been replaced by a version of the Ariane 5 core stage, provided by ATK's partner on the project, EADS Astrium. The Liberty would be able to put [up to] 20,000 kilograms into an ISS orbit, enough to handle any of the crewed vehicle concepts under study.

ATK and Astrium played up the heritage and reliability of Liberty's components while also claiming the vehicle would be cost effective compared to unnamed alternatives. "We will provide unmatched payload performance at a fraction of the cost, and we will launch it from the Kennedy Space Center using facilities that have already been built," said Charlie Precourt, vice president and general manager of ATK Space Launch Systems, in a statement announcing the Liberty proposal. "This approach allows NASA to utilize the investments that have already been made in our nation's ground infrastructure and propulsion systems for the Space Exploration Program."

ATK did not present at the FAA conference last week, but its presence was felt. In his presentation about the CST-100 a day after the ATK announcement, Boeing's John Elbon included an illustration of the Liberty vehicle alongside those of the Atlas 5, Delta 4, and Falcon 9 on a slide emphasizing the capsule's compatibility with various launch systems.

GOVERNMENT WILLINGNESS

Some in industry want to accelerate the pace of commercial crew development. "I think we need to stretch our goals and have commercial crew services operating by 2014," said Michael Gass, CEO of ULA, in a luncheon speech at the FAA conference. That goal would be up to two years earlier than NASA's current projections. He said his urgency came from the need to close the gap in US human access to space with the retirement of the shuttle, forcing NASA to buy seats from Russia, which he called an "embarrassment" for the country. "The commercial crew program should be executed with a sense of urgency, one that the aerospace industry has not seen in decades."

Funding for commercial crew, though, does not match that sense of urgency. While Congress has yet to pass a final 2011 appropriations bill, last year's authorization act included $312 million for commercial crew in 2011 and $500 million each for 2012 and 2013. That's far less than the $3.3 billion that the administration had projected for the program in the same three-year period, but Congressional supporters of the agency have indicated, at least informally, that the full $5.8 billion projected for the program would be made available over six years instead of the original five.

If that money doesn't materialize, though, there could be issues for the commercial crew program. "There are infinite variations" of potential future budgets that make it difficult for NASA to project just how many providers it will be able to support the development of, said Phil McAlister, who is leading the planning for commercial crew program at NASA headquarters, during another panel at the FAA conference. "We would probably have to revisit the fundamental program objectives" [. . .] if the $500-million level was extended beyond 2013, he said. "I'm not saying we would change them, but we would revisit the program objectives."

Drumming up support for initiatives like commercial crew is one of the objectives of a new group. A coalition of fiscal conservatives and free-market advocates unveiled the Competitive Space Task Force last week to convince conservative members of Congress to support the commercial space initiatives of an administration they are usually at ideological odds with. "We're here to try and change the conversation," Rand Simberg, representing the Competitive Enterprise Institute and chairing the task force, said in a Capitol Hill press conference Tuesday.

The task force doesn't have any specific activities planned yet—they said at last week's briefing that they were awaiting details of the administration's 2012 budget proposal—but they indicated they wanted to make clear to conservatives that the administration's commercial crew initiative made better sense than government-run programs and did not, as some critics claimed, jeopardize national security or prestige. "That's what this effort is all about, is to add to our ability to do space, not subtract from our ability to do space," said former congressman Bob Walker.

That's a point companies like SpaceX have also been trying to make. "While this was a significant achievement for SpaceX as a company, we think it's actually more significant as an achievement for the US taxpayer," Hughes said at the FAA conference. He noted the total cost to SpaceX to develop both the Falcon 9 and Dragon was less than $600 million; total company expenditures since its inception in 2002 was $800 million. "When you compare this burn rate relative to other government programs, there is a rather remarkable divide."

NASA leadership has also been making a big push in recent weeks to show its support for commercial crew programs. Earlier this month deputy administrator Lori Garver visited Bigelow Aerospace in Nevada and Sierra Nevada in Colorado, while administrator Charles Bolden witnessed a test firing at NASA's Stennis Space Center last week of the AJ-26 engines that will power Orbital's Taurus 2 rocket.

Bolden also made a surprise appearance at the FAA conference, making a brief speech Wednesday morning where he again emphasized the importance of, and NASA's commitment to, commercial crew providers. "We cannot survive without you," he said. "I can't tell you any stronger. We are big fans of commercial, we are huge fans of commercial space."

"When I retire the space shuttle . . . that's it for NASA access to low Earth orbit," he said. NASA would not compete with commercial providers after that for transporting crews to low Earth orbit, despite Congress authorizing development of the Space Launch System and Multi-Purpose Crew Vehicle as a government alternative to commercial providers. "There's got to be two people in the ring to have a fight. I'm not in the ring for access to low Earth orbit," he said. "We explore."

Today's release of the 2012 budget proposal will kick off a new round of debate about NASA priorities, including for commercial crew; it will also take place in an odd atmosphere where Congress is still working on a final 2011 budget. A bill released late Friday by the House Appropriations Committee would cut $579 million from the agency's original $19-billion budget request, but would give NASA some flexibility to allocate funding within agency accounts.

Despite that debate, Bolden hinted that the debate over commercial crew had already passed. Earlier in the FAA conference, George Nield, associate administrator for commercial space transportation at the FAA, said that while he was optimistic about the prospects of commercial spaceflight, "I'm not sure yet we're at the tipping point today, but we're getting tantalizingly close."

Bolden, in his comments, disagreed with that assessment. "For NASA, that tipping point has long since passed," he said. "We are irrevocably, irretrievably, irreconcilably beyond the tipping point. We can't turn around and go back."

4

A Bold Mission or Sheer Lunacy? Should We Return to the Moon?

Courtesy of NASA Jet Propulsion Laboratory (NASA-JPL)

On December 7, 1992, while en route to the Jupiter system, which it explored from 1995 through 1997, the *Galileo* spacecraft took this photograph of the moon. The distinct bright ray crater at the bottom of the image is the Tycho impact basin. The dark areas are lava-rock-filled impact basins: Oceanus Procellarum (on the left), Mare Imbrium (center left), Mare Serenitatis and Mare Tranquillitatis (center), and Mare Crisium (near the right edge).

Courtesy of NASA

Astronaut Edwin E. "Buzz" Aldrin, lunar module pilot, descends the steps of the Lunar Module (LM) ladder as he prepares to walk on the moon, on July 20, 1969.

Editor's Introduction

The moon has always been a part of the human experience. It causes the levels of our seas to rise and fall. Its periods of waxing and waning helped human beings tens of thousands of years ago develop some of the first calendars—notched bones called tally sticks. The moon influences our art and culture, even our language. The word "month" stems from the old Germanic word for moon, and "lunacy" and "lunatic" derive from *Luna*, the Latin word for moon, because ancients believed full moons caused insanity. In the decades before the first lunar landing in 1969, young people were often told to "shoot for the moon." Today, however, the moon seems less an object to aim for than one to get past in order to accomplish greater feats.

In announcing his new plan for NASA, President Obama rejected a return to the moon, noting that "we've been there before." Disappointed NASA employees who had been working on the Constellation plan, which would have returned Americans to the moon, aren't the only ones critical of NASA's new direction. Many wonder if, after just six manned visits, mankind has really learned all it can from the moon, a place the Apollo astronauts had only just begun exploring. Still, over the course of their half-dozen trips, U.S. astronauts made significant findings, discovering that the moon is geologically related to the Earth; that it is not a primordial part of the solar system, but rather a planetoid that evolved over time; and that despite this evolution, it is a preserved part of the early solar system that grants us a better understanding of what the terrestrial planets—Mercury, Venus, Earth, and Mars—may have once been like. Robotic probes sent by various nations have since confirmed the existence of water on the lunar surface, as well as the prevalence in the lunar regolith of helium-3, a stable isotope that may serve as a clean fuel in nuclear fusion.

Those opposed to sending additional astronauts to the moon point to these discoveries, made by unmanned orbiters and landers, to argue that it is unnecessary to risk human lives in lunar exploration. The moon has always been lifeless and is incapable of supporting life, they point out; additional lunar landings will not prepare us to go to Mars, which may have actually once supported life. Further, some argue, the Red Planet could be a candidate for terraforming, a theoretical process by which man would alter the atmosphere, temperature, and topography, prepping the planet for human colonization. Supporters of Obama's plan contend that a manned mission to Mars would be a more effective use of astronauts' talents. Owing to the vast distance between Earth and Mars, astronauts would stay there longer than they would on the moon, and their protracted stays would require more hands-on research.

Although the question has, from a government-funding standpoint, been answered—at least for the time being—the articles in this chapter ask if another round of lunar exploration would be worthwhile. In the first article, "Good Night, Moon?" taken from *Current Events*, the author briefly details both sides of this contentious issue. Jeffrey Marlow continues the conversation with "Moon-Rush: Is the United States Sitting Out of the Next Space Race?" and explains how China, Japan, and India are using the moon as a proving ground for their budding unmanned space programs. These nations are also in the initial stages of planning manned missions to the lunar surface. Next, in an interview with Brittany Sauser for *Technology Review*, pioneering astronaut Buzz Aldrin presents his arguments against returning to the moon and outlines a plan for reaching Mars in the coming decades. Finally, in a piece for *Ad Astra*, Gordon Woodcock draws on his more than a half-century of experience in aerospace engineering and mission analysis to discuss the economic benefits of returning to the moon and going on to Mars.

Good Night, Moon?*

Current Events, February 22, 2010

Some say it's a not-so-small step forward for the future of space exploration. Others argue that it's a giant leap backward for NASA, the U.S. space agency. U.S. President Barack Obama's latest budget proposal calls for the cancellation of the Constellation program, which is developing spacecraft to launch astronauts back to the moon. Canceling the program would put the future of sending humans to the moon in jeopardy.

The Constellation program is part of a bold U.S. space strategy that would send astronauts to the moon by 2020 and possibly build a permanent base there. Between 1969 and 1972, six American Apollo spacecraft landed on the moon, and 12 astronauts explored the moon's surface. Yet no one has set foot on the lunar landscape since.

President Obama's proposal would shift billions of dollars to other outer space endeavors. His budget includes plans to fund commercial spacecraft to carry NASA astronauts into space. The policy would also support the development of systems to build larger, more powerful spacecraft that might push the limits of space exploration.

TO THE MOON!

Many people say that scrapping lunar missions would be a moon-sized mistake. "It has taken over 50 years to build and develop America's ascension to its rightful place as the dominant player in human spaceflight," says U.S. Rep. Pete Olson (R-Texas). "That dominance is apparently no longer desired." Critics say that ending the Constellation program will enable China and Russia to eclipse the U.S. as leaders in space exploration.

Many experts argue that we still have a lot to learn about Earth's closest neighbor. For example, last fall NASA found evidence of water ice below the moon's surface.

"We now have confirmation that sites that Apollo visited were not representative of the entire moon," says Clive Neal, chairperson of NASA's Lunar Exploration Analysis Group. "It is like sending six landers to California and saying we now know everything there is to know about North America." Further exploration of the moon may unlock clues about our solar system.

Neal and others point out that new moon missions could yield other breakthroughs that improve life on Earth. Home insulation, medical scanning techniques, and freeze-dried food are just a few of the new technologies developed by the Apollo program in the 1960s.

SO OVER THE MOON!

To many space experts, sending astronauts to the moon represents the past, not the future, of space exploration. Last year, Buzz Aldrin wrote in *Popular Mechanics* that any plans to revisit the moon are "a glorified rehash of what we did 40 years ago." On July 20, 1969, Aldrin became just the second person to set foot on the moon. He is among the experts calling for the U.S. to turn its attention to exploring Mars.

Experts also point out that NASA has continued to push the envelope since it ended moon landings in 1972. The Mars rovers and the Hubble Space Telescope are only two of the cutting-edge programs created by the space agency in the past few decades. Supporters say that such innovation can continue without expensive human trips to the moon. In its four years, the Constellation program has cost taxpayers about $9 billion. That's a lot of money for few results, according to the program's critics. They also note that the program is behind schedule and the technology is becoming outdated.

Many people say that the new vision for NASA will make the U.S. stronger, not weaker, in the space race. NASA administrator Charles Bolden, a former astronaut, calls it "a 21st-century program of human space exploration" that will "push the frontier of [space] research."

Moon-Rush*

Is the United States Sitting Out of the Next Space Race?

By Jeffrey Marlow
Ad Astra, Spring 2009

In the early morning dawn of October 22, 2008, the sky over southeastern India was illuminated by the fiery exhaust from the Chandrayaan-1 spacecraft, which was embarking on India's first robotic mission to the Moon. Two weeks later, Chandrayaan became the third orbiter actively studying the Moon: mapping the surface, characterizing the mineralogy, and searching for traces of water. This accomplishment may not seem particularly surprising given the success of recent autonomous missions to Mars and other planets; but what *is* noteworthy is the list of countries participating in this modern-day Moon-rush. Asian upstarts China, Japan, and India have taken center stage, while traditional space powers such as the United States, Russia, and the European Space Agency have been noticeably absent.

India's recent success, coupled with China's first spacewalk in September, signals the acceleration of a new Asian space race—a technological arms race that is quickly closing the gap between the United States and the rest of the world. As China, India, and Japan push each other to aim ever higher, NASA has been relegated to the sidelines, nursing a broken shuttle fleet and a strained budget.

The Asian space race was ignited in large part by China, which in 2003 became just the third nation to achieve manned spaceflight. China's manned missions—there have now been three—are all part of a long-term agenda that will see the construction of a Chinese space laboratory and permanent space station. The program, titled Project 921, has been on the books since 1992, reflecting the persistence and foresight with which China has pursued its spacefaring ambitions.

Though it has been a long time in the making, China's space program is finally achieving visible results, and the rest of the world has noticed. China's distinctive blend of militaristic tendencies and nationalistic fervor, highlighted by a 2007 anti-satellite missile test, has triggered a powerful reaction. Other Asian powers have responded, and a new story has begun to unfold in the regional battle for cultural, technological, and economic supremacy.

Japan flinched. "We were surprised," flight engineer Masashi Okada told *Time* magazine soon after China's historic mission. "Obviously we knew they were working toward it, but they achieved manned flight very quickly. We are fully aware that our space development program has to include manned spacecraft." Of the current Asian contenders, Japan has the most impressive pedigree of space exploration, having pioneered important astronomical tools, constructed reliable launch vehicles, and helped build the International Space Station. But in 2003, encouraged in part by the perceived threat of China's fledgling space program, Japan rebranded its agency and focused its effort not on international partnerships but independent missions. Japan hopes to develop its own manned spaceflight program and land astronauts on the Moon by 2020.

India also took drastic action following China's 2003 manned mission. Since the mid-1960s, the Indian Space Research Organization (ISRO) has been content to use space technology for development programs like communications, health care, and environmental monitoring. But in 2006, its national pride piqued, India announced more ambitious plans, calling for a manned program that would allow it to assume its rightful place as a technologically progressive nation. In the wake of Chandrayaan's launch, Indian officials were more direct than ever. "China has gone earlier," ISRO satellite communications director Bhaskar Narayan told Reuters, "but today we are trying to catch them." India's manned spaceflight plans remain unspecified, but there is little doubt that ambitions run high. Some analysts predict that the first Indian astronauts will fly around 2014, and the ISRO maintains that India's ultimate goal is a 2020 Moon landing.

Unfortunately, as the Asian space race heats up, NASA seems to be losing momentum. The space shuttles, which may have been futuristic enough 30 years ago, are now antiques. It has become increasingly difficult to justify the safety risks and billion-dollar price tags that come with each launch. Wisely, the remaining shuttles are on the fast track to retirement, but thanks to cost overruns, leadership gaps, and sagging employee morale chronicled by a NASA safety panel in August, a replacement vehicle is still several years off. In the interim, from 2010 to 2015, American astronauts will have to hitch rides to the International Space Station with the temperamental Russians, which is kind of like having to ask your grouchy neighbor to let you into your own house. Once things do get back on track, NASA aims for a manned return to the Moon by—you guessed it—2020. But even that might be too late: Former NASA Administrator Michael Griffin recently admitted the possibility, telling the BBC that "if China wants to put people on the Moon, and if it wishes to do so before the United States, it certainly can."

So what are Americans, seemingly the laggards in the modern space race, to make of this Asian space enthusiasm? Does the shifting momentum signal a new world order, or is the contest an inconsequential rerun playing 40 years after the original? Recent patterns notwithstanding, the United States retains the world's most advanced space program, thanks to the largest space exploration budget in the world and several decades of persistent, diligent, sustained effort. NASA has unparalleled technical and administrative know-how that can't be bought off the shelf, and though the manned space program is faltering, robotic missions are returning troves of scientific data from across the universe.

But the trends do matter, and the future of America's space program may have a monumental effect on the future of the entire nation. The nebulous sense of prestige gained from a strong space exploration program often translates into substantive economic and political gains. James Oberg, a Houston-based space consultant, told the Associated Press that "doing 'Moon probes' advertises a country's technological level, and that's good for high-tech exports, and for validating the threat-level of its high-tech weapons."

Space exploration is also a bellwether for a country's philosophical outlook. A nation excitedly engaged in space exploration is a nation that believes in its future, one that tackles new problems confidently and leads fearlessly. It is a nation where schoolchildren idolize astronauts and rocket scientists like most youths today admire football stars—an attitude that eventually develops into a wide base of scientific and technological knowledge, driving innovation and economic growth. It is a nation that is going somewhere.

A nation with a stagnating space program, on the other hand, is one on its heels, stricken by intellectual malaise. It is a nation that has turned inward and is unwilling or unable to take risks that are likely to provide economic and societal rewards. It is a nation that has lost its hunger and grown a little too comfortable in its privileged position, happy to reflect on past glory days.

The exact date that China, India, Japan, the United States, or anybody else next lands on the Moon is immaterial, but the race itself is important as a litmus test of participating nations' technological prowess and the value they ascribe to science and technology. Right now, the United States must work to get back on the right side of the equation. Meanwhile, half a world away—after Chandrayaan soared over rice paddies and fishing ships toward the Moon just a few months ago—India is looking up.

Buzz Aldrin*

Apollo Astronaut Says: Forget the Moon, Let's Colonize Mars

Buzz Aldrin with Brittany Sauser
Technology Review, July/August 2010

In April, President Obama flew to Kennedy Space Center in Cape Canaveral, FL, to reveal details of his new strategy for NASA and the future of U.S. spaceflight. Sitting next to the president on Air Force One was Buzz Aldrin, who in July 1969 became the second man to walk on the moon. The seating arrangement was appropriate, since both men share a common goal for the nation's space program: reaching Mars by the mid-2030s.

Like Obama, Aldrin opposes the strategy set by President Bush in 2004 to return humans to the lunar surface by 2020. The cornerstone of Bush's plan for NASA was the Constellation program, which included building two new rockets—Ares I to ferry humans into orbit and Ares V to transport heavy cargo—and a manned exploration vehicle called Orion. But the program fell behind schedule and was over budget *(see "The Future of Human Spaceflight," January/February 2010 and at technologyreview.com)*. In January, Obama released a new budget proposal that increased NASA's budget by $6 billion over the next five years but terminated the Constellation program.

Technology Review reporter Brittany Sauser recently asked Aldrin about his ideas for the future of U.S. human spaceflight.

***TR*: Why not go back to the moon?**

Aldrin: We explored the moon 40 years ago, and now it should be developed by robots for scientific, commercial, and security reasons. Basically, I don't see a financial return to justify the cost of sending U.S. humans and rockets back to the moon; it's a waste of decades and hundreds of billions of dollars.

What should NASA focus on instead?

The objective should be a permanent presence on Mars by 2035. That's 66 years after Neil Armstrong and I first landed on the moon, and our landing was 66 years after the Wright brothers' first flight. Mars is clearly the best permanent-residence location other than Earth, and we can go there in case somebody or something blows up Earth. We will have a place that ensures the survival of the human race. That means humans who go there commit to staying—one-way tickets will be technically easier and less expensive and get us there sooner.

But that will take years. What should NASA's transition strategy be?

Ares I and Ares V should be canceled, and in their place we [should] build an evolutionary shuttle-replacement launch system that could be called something like Ares III [and would transport both people and heavy cargo]. Orion should continue to be developed as an emergency vehicle for the space station, as the president stated. Meanwhile, I also very strongly suggest that instead of retiring the shuttles [in late 2010] and buying rides with the Russians for five, six, or seven years to get to our $100 billion space station, a highly undesirable situation, we stretch out the flights of the five remaining shuttle orbiters to 2015.

The president's plan also relies heavily on the commercial space industry to provide crew and cargo transportation to the space station. Do you think that's a good idea?

Yes, I do. Commercial vehicles will also help fill the gap, so we can develop new launch vehicles and spacecraft for landing on runways for the years beyond 2015 to bring us to the threshold of Mars.

How will we get to Mars by 2035?

We build the ultimate transportation system: a cycling spaceship called the Aldrin Cycler, which I first unveiled in 1985. It cycles between Earth and Mars. Spacecraft can hook up to it, and we could use it to fly by a comet as early as 2018. Then, in 2020, we could travel to a near-Earth object. We would need to build in-flight refueling and communication relays before and during this time, with more visits to asteroids.

In 2025, we land unmanned on Phobos [a moon of Mars] with some elements of habitation. We land people there in 2027 for a year and a half; in 2029 for a year and a half; and in 2031, we land three people who will not come back. In 2031, six people coming from Earth will join the three at Phobos and then continue on to become the first people to land on Mars by 2033 or 2035.

But a consensus on Mars as the goal destination has not been reached. Have you spoken with other influential Apollo astronauts who oppose the termination of Constellation?

I have long been open to discussions with other astronauts, especially the 24 astronauts, 18 of whom are still alive, who reached the moon. But that exclusive group does not have any coherent organization. I am forming a nebulous but much-needed concept for an organization that I call the Unified Strategic Space

Enterprise. It would consist of highly respected people who would assist in the development of the national space policy.

It seems like we've been arguing about the future of the U.S. space program for decades now.

We really have only been debating the human spaceflight portions of exploration; where do we send U.S. humans? But there are robotics, the space station, technology developments like in-flight refueling, and all sorts of other things to think about.

The Moon and Mars—Stepping Stones to Life in Space*

By Gordon Woodcock
Ad Astra, Winter 2009/2010

NASA's approach to human exploration of space is rooted in the Apollo program. Apollo was a vital part of the Cold War. It had the highest national priority, with the goal of demonstrating that the United States was the world's most capable space power. Young people were highly motivated by progress in space development and the idea that their future could lie in space—not only as space engineers, but as astronauts, space workers, or settlers. During the Apollo program, space prowess was the premier indicator of scientific and technological leadership.

But today, the Cold War is over, and an Apollo-like approach to space exploration no longer satisfies our needs. *The Economist* (January 2009) writes, "In space travel, as in politics, domestic policy should usually trump grandiose foreign adventures. Moreover, cash is short and space travel costly. Luckily, technology means that man can explore both the Moon and Mars more fully without going there himself. Robots are better . . . They can also be made sterile . . . which humans cannot."

If the main reason for human missions is science, *The Economist*'s view is right on. There is another view, which John Marburger put succinctly: "It's about bringing the inner solar system into our economic sphere."

If economic gain drives space exploration, that places tough constraints on human missions; cost must be less than the value of benefits. Estimates of feasible cost indicate the current approach for putting humans on the Moon is about 50 times too expensive. Here's a recipe for dramatic cost reductions:

- Major cost reductions can be achieved in transportation through the use of long-life reusable systems and lunar propellants.
- Even then, frequent transporting of people back and forth from the Moon is unaffordable; tours of duty of at least several years are needed.

- If people stay several years, they must be allowed to travel with immediate family, who also work for the enterprise.
- Transporting food is unaffordable, so bioregenerative life support systems with food production are essential. Carbon and nitrogen biomass inventory may initially come from Earth and be recycled. These substances are available on the Moon only in parts per million and may not be a practical byproduct of industrial operations. Hydrogen and oxygen, however, are available.
- Transporting infrastructure (habitats, utilities, industrial equipment) is unaffordable for more than a dozen or so people, so most infrastructure must be made on the Moon.
- Space settlement, therefore, is not an option; it is an imperative if we want humans beyond low Earth orbit to benefit human civilization.

An example scenario begins with a six-person, six-month mission to the Moon using reusable transportation. Lunar oxygen production would need to be installed first, perhaps robotically, to support the astronauts and provide propellant for a lunar transporter. The transporter is first used in cargo mode to get flight experience. To reduce costs, lunar resources are essential early, first for propellant and later for structure manufacturing. A reusable launch vehicle is introduced to further reduce cost. If experimental lunar food growth is successful, the settlement may be expanded to 12 people serving one-year tours. Then lunar infrastructure production begins, with the goal of manufacturing 30 percent of food growth modules and 20 percent of habitation systems on the Moon and doubling the settlement to 24 people serving two-year tours. As infrastructure and production continue to grow, the population also grows from dozens to hundreds of people living on the Moon for longer and longer tours. The growing population is supported almost entirely through increased self-sufficiency.

New technologies will reduce the cost of Mars missions so that settlements are possible there too. But the Moon and Mars cannot support as many inhabitants as Earth. Continued growth demands efficient use of resources in new ways. Gerard O'Neill recognized that in the 1970s. It was one motivation for his free-floating fabricated space habitats. Resources of the inner solar system are sufficient to support hundreds of times Earth's population in such solar-powered habitats. For example, a conventional habitable spacecraft such as ISS provides about 100 m^3 of atmosphere per person. An O'Neill habitat provides about 10,000 m^3 per person. Earth's total atmosphere is roughly 10,000 times *that*, taken as volume at sea level pressure.

Structure for an O'Neill habitat would be mostly steel. There are enough nickel iron asteroids in the main belt to supply steel for enormous numbers of habitats. Nitrogen to fill them may be the limiting factor (oxygen can be extracted from rocks). Icy cometary objects or outer planet atmospheres are potential sources.

One problem is the productive effort required to turn resources into fabricated ecosystems. An O'Neill habitat represents a thousand times more manufactured hardware per person than that of modern industrial societies. (A lunar or Mars settlement only represents a few times more hardware.) The solution is highly automated and robotic manufacturing from raw materials through finished product. Lunar and Mars settlements will start this trend.

GORDON WOODCOCK *is former president of the L5 Society and former chairman of the Executive Committee of NSS, with more than 50 years in aerospace engineering and mission analysis.*

5

Greenlighting the Red Planet: Should We Go Straight to Mars?

Courtesy of NASA/USGS

This mosaic of Mars is a compilation of images captured by the Viking Orbiter 1. The center of the scene shows the entire Valles Marineris canyon system, over 3,000 km long and up to 8 km deep, extending from Noctis Labyrinthus, the arcuate system of graben to the west, to the chaotic terrain to the east.

Courtesy of NASA Jet Propulsion Laboratory (NASA-JPL)

An artist's rendering of the Mars Exploration Rover.

Editor's Introduction

Like the moon, Mars has long gripped the human imagination. It is the only planet in the solar system that can be seen in detail from Earth, and at its brightest, it outshines every planet in the night sky, save for Venus. Peering through their telescopes, astronomers such as Giovanni Schiaparelli and Percival Lowell became convinced they saw a network of canals on Mars and theorized that an ancient civilization had built them in order to bring water across the planet from its polar ice caps.

Talk of intelligent life on Mars inspired many authors in the late 19th and early 20th centuries to imagine what kinds of civilizations might live on the blood-red planet. In his 1898 novel *The War of the Worlds*, H.G. Wells describes an advanced but hostile race of Martians intent on conquering the Earth. In his Barsoom series, Edgar Rice Burroughs writes of a frontier-like Mars, on which his main character, a Civil War veteran named John Carter, must use his skills as a soldier in order to survive. In his 1950 short story collection *The Martian Chronicles*, Ray Bradbury dreams of a Mars populated by human beings after the original telepathic inhabitants have perished.

These and other romantic images of the fourth planet in our solar system were put to rest in July 1965, when the U.S. probe *Mariner 4* relayed photos of a Mars that was dry, barren, and very likely incapable of having supported life at any point in its recent history. In reality, Mars was nothing like what humans had imagined. That said, dozens of subsequent unmanned orbiters, landers, and rovers have revealed a remarkable world, mysterious and tantalizing. Despite its overt differences, Mars shares similarities with Earth.

While mankind has much to learn about the Red Planet, here's what we know for sure: Mars' diameter is roughly 4,200 miles, a little more than half that of Earth's. Mars orbits our sun every 687 Earth days, meaning Martian years are almost twice as long as ours. Days on Mars are longer, too, though only by about 40 minutes. Its thin atmosphere consists of mostly carbon dioxide, a compound that is absorbed by terrestrial plants, which exhale oxygen, one of the key components of our atmosphere. As on our planet, the north and south poles of Mars are covered in ice that grows thicker during its winters. The planet's reddish color is due to the abundance of rusty iron in its soil. (Iron on Mars rusts just as it does on Earth.) Mars is home to the largest known volcano in the solar system, Olympus

Mons, which, at 15 miles high, is three times taller than Mount Everest, the highest mountain on our planet. We also now know that water once flowed on the surface of Mars and may still do so today.

Here's what we don't know: Did Mars ever harbor life—even the most basic microbes? Is Mars as it exists now a look into the future of our own blue-green Earth? Is it possible for human beings to survive for extended periods on Mars? Would it be possible to make Mars more habitable through terraforming?

The desire to answer these types of questions has led many to call for manned exploration of Mars. Robotic surveys are well and good, advocates insist, but machines can only move as far as earthbound controllers tell them to go—and slowly at that. What's more, unmanned missions can get bogged down or lost; in fact, two-thirds of such missions to Mars have failed for unknown reasons. Even if probes are capable of surveying most of the planet, some have argued, it will take human beings, working with reason and instinct, to truly uncover its many secrets.

But other questions remain, such as whether it's feasible to mount such a mission using our current level of technology. Even if it is possible, it is worth the considerable cost—measured in dollars and potentially human lives? Is President Barack Obama's controversial program—one without specific dates or targets—the best way to reach Mars? The selections collected in this section seek to address some of these queries. In the chapter opener, "Destination Mars," written for *Maclean's*, Kate Lunau details the ways in which humans have already begun preparing to go to Mars. In the next entry, written around the time President George W. Bush announced Project Constellation, Gregg Easterbrook discusses the enormous costs and the tremendous technical hurdles that would face any manned mission to Mars.

In the next piece, Sam Howe Verhovek describes a radical new nuclear-plasma propulsion drive being developed by Franklin Chang Díaz that could get astronauts to the Red Planet in just 39 days. (Using conventional chemical rockets, such a trip would take approximately six months.) In the following selection, Brian Palmer highlights some of NASA's new ideas for propelling Mars-bound spacecraft, including one that involves solar power. Finally, in "Making Mars the New Earth," Robert Kunzig explains how Mars could be terraformed with present technology.

Destination Mars*

By Kate Lunau
Maclean's, September 27, 2010

Viewed through a telescope on a clear night, the planet Mars glows a soft, dullish red. It seems foreign and strange, but familiar, too: like Earth, Mars has polar ice caps, clouds drifting in its thin atmosphere (even snow), and changing seasons. Its day is just 40 minutes longer than our own. And even though it's now a freeze-dried wasteland, a growing body of evidence suggests Mars was once wet and warm, and might have harboured life around the same time life sprung up here. Human explorers are bound to set foot on Mars one day. And it might be sooner than most of us think.

But our neighbouring planet, fourth from the sun, is also unimaginably remote: at its closest point in orbit to Earth, which happens only once every 26 months or so, Mars is still about 200 times farther away than the moon. At best, it would take a manned spacecraft roughly six months to reach it. By comparison, "the moon is three days away," says Bret Drake, who leads mission planning and analysis for the Constellation Program at NASA's Johnson Space Center. "You can go any time, and if things go wrong, you can return any time." Once a spaceship left Earth's orbit for Mars, there'd be no turning back.

On the surface, astronauts might have to contend with everything from swirling dust storms to blasts of radiation from powerful cosmic rays. Their research would be a scientific bonanza, teaching us about our solar system, about the genesis of life on Earth and maybe even whether life exists on Mars, or ever did. Observing how the crew's bodies change in reduced Martian gravity could tell us if it's really possible to survive for years on end in space. They'd have to wait over a year until the planets lined up to come back, making it a 2½ year trip, all told. If something went seriously wrong, there'd be little to no hope of rescue.

Teams of scientists and specialists from around the world are already working on projects that tackle some of the biggest challenges of a Mars mission, changing the way we think about space travel, about human endurance and about how

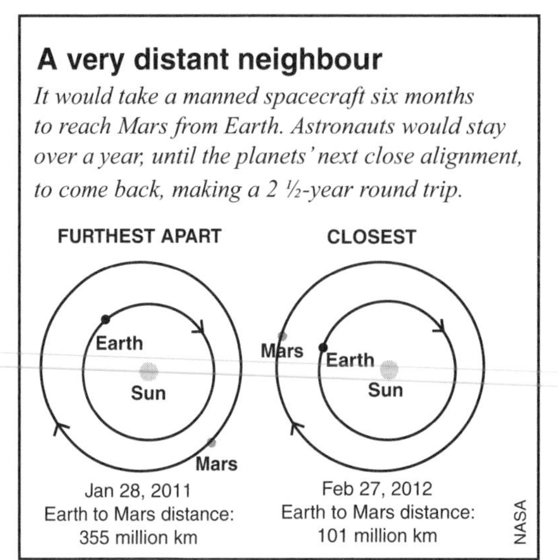

we might live someday beyond the bonds of Earth. At the Kennedy Space Center in April, U.S. President Barack Obama delivered a historic speech on space exploration. "We'll start by sending astronauts to an asteroid for the first time in history," he said. "By the mid-2030s, I believe we can send humans to orbit Mars and return them safely to Earth. And a landing on Mars will follow. And I expect to be around to see it."

Space-age luminaries like Buzz Aldrin—who, along with Neil Armstrong, was the first human to set foot on the moon, in 1969—call it our next frontier. "Mars is the only other place that approaches conditions here," Aldrin, 80, told *Maclean's*. "It's much closer to Earth than Venus or Mercury," the only other rocky planets in our solar system. Unlike other destinations, "you can imagine astronauts on the surface of Mars, moving and working," says Richard Zurek, chief scientist for the Mars Program Office at NASA's Jet Propulsion Laboratory. "I truly believe the Mars astronauts are alive today," says Canadian astronaut Robert Thirsk. "They're probably in elementary school right now."

In June, six men entered a sealed isolation chamber in the outskirts of Moscow, to remain there for 520 days. The Mars500 study, a joint effort of the European Space Agency and the Russian Institute for Biomedical Problems, is an attempt to recreate the mental and physical stresses of long-duration space travel, and the effects of extreme isolation. (These lessons also come in handy on Earth: NASA is lending its expertise to help 33 Chilean miners now trapped below ground, expected to be there for up to four months until rescuers can reach them.)

On their simulated mission to Mars, the men—three Russians, two Europeans and one Chinese—only have personal contact with each other. A 20-minute delay is built into communications with the control centre, the same length of time it takes for messages to travel one way between Mars and Earth. The habitat's main living quarters, where each man has his own tiny bunk, is just 3.6 by 20 m.

In one diary entry in July, French engineer Romain Charles wrote about spending his 32nd birthday in isolation. On his last visit home, "I received some presents for my first birthday in the modules, for [Christmas] and also for my next birthday in 2011," he writes. "Now here I am with a lot of gifts just under my bed and nothing to stop me from opening them." Entering the living room, he found his crewmate, Italian-Colombian engineer Diego Urbina. As a surprise, Charles recounts, Urbina had taken a photo of an astronaut, "changed his face to mine and the flag for the French one," and asked all the crew to sign it. "He knew that since

I was nine or 10 years old I wanted to go to space and he made this dream come true in a way."

Like the Mars500 group, the first team to go to Mars might have four to six members, Drake says, and a complementary set of skills: a commander, an engineer, a geologist and a doctor is a likely mix. They'd almost certainly be multinational and include both men and women. And we can only hope they get along as well as the Mars500 team apparently does. "It's going to be a very isolated spacecraft, away from family and friends," says Thirsk, who spent six months aboard the International Space Station last year, becoming the first Canadian to fly a long-duration mission there. Thirsk could often speak to his family back home, a luxury they won't have.

If there's friction among the crew, reaching a mediator might not be possible either. Even when psychologists are available, astronauts "often try to hide emotional problems, out of fear they'll be grounded," Mary Roach writes in her new book, *Packing for Mars*, and stressed-out astronauts have been known to vent their frustrations at mission control. Conflict resolution software, in which "the computer acts as a therapist," might be helpful, says Dr. Jeffrey Sutton, director of the National Space Biomedical Research Institute. It would give astronauts the chance to play out conflicts—hypothetical or real—and explore outcomes with a machine, instead of on a crewmate.

When Thirsk was on the ISS, he spent long moments gazing down at Earth. "I was amazed by its beauty," he says. "The oceans are blue, but they're 100 shades of blue. You see incredible patterns in the desert: 100 shades of brown, gold and red. It's so heartwarming to see such a beautiful planet, and all the signs of life down there." This is common among astronauts, who tend to say that seeing Earth is the greatest benefit of their time in space, says Dr. Nick Kanas, a professor at the University of California, San Francisco and an expert in astronaut psychology. It can be calming and restorative, he says, imparting "a sense of history, of a lack of boundaries, and of the beauty of Earth as a homeland."

Astronauts going to Mars won't have that benefit. They'll be the first humans to see their home planet fade away, until it disappears into the blackness of space. (As they zoom toward Mars through permanent sunlight, they won't even see any stars, Roach reports, just black.) "Nobody in the history of our existence has ever perceived Earth as an insignificant dot. We've either seen it as a beautiful ball, or we're standing on it," Kanas says. Nobody knows what the impact of "Earth out-of-view phenomenon" will be. "It might be nothing," he says, "but it might be profound."

Watching their home planet disappear out the rear-view window is just one of the mental and physical challenges these astronauts will face. Microgravity, stress and radiation all wreak havoc on the human body; in space, immune systems become weaker and wounds take longer to heal, Sutton says. Radiation can make medication less effective, and in reduced gravity, muscle and bone tissue wither away. Astronauts might lose bone at up to 1.5 per cent per month on average, a loss that "can be as high as 10 times that of post-menopausal women," he says.

After months in space, an astronaut returning from Mars might step back into Earth's gravity and snap a bone. Researchers are looking at drugs to help combat bone loss (Thirsk was part of a drug trial on the ISS). Others talk about designing a spinning spacecraft, to create artificial gravity. For now, "the main countermeasure," Thirsk says, "is exercise." On the ISS, astronauts do resistance training and work out on stationary bikes or treadmills, with a bungee cord to keep them from floating away.

In May, astronaut Chris Hadfield, who'll be the first Canadian to command the ISS when he takes over in 2012, spent two weeks in a pod under water off the Florida coast. On that NASA mission, NEEMO 14, he and three crewmates simulated activities that might be performed on Mars or the moon, including emergency procedures like rescuing a wounded colleague (played by a dummy) from the hostile environment of the ocean floor. They also charted how their bodies reacted to the confined space. "It's very evocative of being on another planet," Hadfield told *Maclean's* in a video chat from the pod. "When you walk outside, you're weighed down like on another planet. You bounce when you walk, and you hear every breath."

If there were an emergency and surgery were required, a robot might be the one to do it. McMaster University's Dr. Mehran Anvari is an expert in telerobotic surgery, remotely controlling robotic arms that perform operations hundreds of kilometres away. The system was designed for use in remote communities, but Anvari's worked with the Canadian Space Agency and NASA—including mock-ups on NEEMO missions—to see if it could be done in space. Because of the long delay in communications to Mars, "you'll need a system with some degree of artificial intelligence," says Anvari, who directs the Centre for Minimal Access Surgery. "I'll be very surprised if the next long-term manned mission includes a surgeon," he says. "It's not necessary."

A robotic surgeon would certainly be an asset on a mission to Mars, but we still don't have a ship that can take us there. With current technology, a trip would require a heavy-lift rocket (one that could blast at least 70 metric tonnes off Earth's surface) and several launches to get all the necessary equipment off the ground. The giant ship would then be assembled in orbit before blasting off to Mars.

Lifting hundreds of thousands of pounds into Earth's orbit is no small feat. Paul Spudis, senior staff scientist at the Lunar and Planetary Institute, suggests it might be possible to reduce weight—and, critically, launch costs—by making rocket fuel on the moon instead of bringing it all from home. "The moon's poles have significant amounts of water ice," he says. This could be broken down into hydrogen and oxygen, "the most powerful rocket propellants we know of." (Obama has said he hopes to choose a heavy-lift rocket design by 2015.)

Former NASA astronaut Franklin Chang Díaz thinks it's time to move on from chemical rockets. After retiring from NASA in 2005, he founded Ad Astra Rocket Company, which is developing a plasma rocket called VASIMR. Powered by solar arrays or nuclear electric power to produce a superheated gas (plasma), "we can do the same work with about one-twentieth the fuel," he says. "It completely changes the equation of space travel." With energy close to what's generated in a nuclear sub, he believes, VASIMR could bring humans to Mars in 39 days. After that, "We won't stop at Mars," he says. "We'll keep going."

After months of travelling through space, the ship will park itself in orbit around Mars. A lander will detach, ferrying the crew down to the planet's surface, where—if the most commonly expected scenario plays out—they'll find rovers, supplies and a habitat waiting, delivered robotically on earlier missions.

Astronauts will spend days at a time inside a rover, exploring the surface and collecting samples, says Kriss Kennedy, a space architect at NASA. (The pressurized rover, which might be the size of a Winnebago, would be what astronauts call a "shirt-sleeve environment.") In reduced Martian gravity, which is three-eighths of our own, they won't need to be strapped to their beds when they're sleeping; but they will feel like they have superhuman strength, jumping higher and lifting more than they ever could on Earth.

When not in the field, astronauts will come back to a habitat, which might be buried in the ground for protection against radiation that bombards the planet. Its size will depend on the specifics of the mission, but each crew member would probably have his own private quarters, with shared space for exercise, lab work and socializing. (The Mars500 crew kills time playing Guitar Hero.) The air they breathe could be brought along in canisters, or made from waste water split into oxygen and hydrogen, a process used on the space station.

It's impossible to really imagine setting foot on another planet, but luckily we can practise at home. The spot on Earth that most closely resembles Mars is here in Canada: Devon Island, in Nunavut, which was struck by a meteorite about 39 million years ago, leaving a massive scar called the Haughton Crater. Today, it's "an almost perfect place to learn how to explore Mars," says Pascal Lee, a planetary scientist at NASA Ames Research Center and chairman of the Mars Institute, which runs the Haughton-Mars Project, a scientific study of the crater and how it relates to Mars. "Mars is a cold, windy, dusty, barren and impact-scarred place," he says. "So is Devon Island."

This summer, the Haughton-Mars Project team spent six weeks on site, its 14th consecutive season. Part of their research involved trying out a suitport, a system in which a spacesuit is fastened to the outside of the spacecraft (conventional suits are stored inside the ship and donned inside an airlock). When an astronaut wants to do a spacewalk, he swings open a port "like a thick refrigerator door," Lee says, climbs into the suit and "scoots through legs first." He then seals the port behind him, detaches from the vehicle and walks off wearing the suit. When the spacewalk is over, he re-docks, climbs out of the suit and right back into the ship. Suitports would have several advantages on Mars: they're quick to put on, and corrosive Martian dust is kept outside, although they do risk damage from exposure. Up in Nunavut, the team experimented with spacesuits dangling from a Humvee, which stood in for the type of rover that astronauts could drive on Mars. During the exercise, "almost the entire team tried going in and out," Lee says. "It was fantastic."

On Mars, astronauts will find mountains taller than Everest, canyons deeper than the Grand Canyon, and the biggest volcanoes in our solar system: the largest, says Zurek, is the size of the state of Arizona at its base. They might encounter massive winds and dust storms that can engulf the entire planet, and temperatures from -125° C at the poles in wintertime, up to 20° C at noon by the equator, according to NASA. And it might not always look so different from home. In 2008, aided by a Canadian-made weather station, NASA's Phoenix Mars lander found the first evidence of snow falling from Martian clouds.

Space agencies have been sending robotic missions to Mars since the 1960s, but human explorers would achieve more than robots ever could. Spirit and Opportunity, two NASA rovers, landed on Mars in 2004, and while Spirit eventually got stuck in a sand trap, Opportunity's still driving around. But their work is excruciatingly slow: it's been said humans could learn as much in one week on the ground. Getting samples into scientists' hands will be invaluable. Unlike Earth, the Martian surface seems to be one solid crust; with no plate tectonics to melt it down, ancient rock litters the ground, "preserving an early record of the planet," Zurek says. Beyond just teaching us about how Mars has evolved, this record might teach us about the origin of life.

Today, the pressure on Mars is so low that liquid water boils off almost immediately. But scientists have found evidence the planet was once wet and warm, with lakes, oceans, maybe even an atmosphere dense enough to support rain. Where there's water, there can be life, and new clues suggest that maybe it's still there. We now know there's methane in the Martian atmosphere, which, in Earth's atmosphere, "is largely produced by biological matter, like plants and algae," says Victoria Hipkin, senior planetary scientist at the CSA. "Methane in the Martian atmosphere could mean there's a surface source," maybe even methane-producing microbes, "some of the most primitive microbes that exist on Earth."

On Axel Heiberg Island, in Canada's Far North, a team of scientists has been exploring the remote Lost Hammer spring. Its water pushes up through 600 m of permafrost, says McGill University microbiologist Lyle Whyte, but it's so salty it doesn't freeze. The water lacks any consumable oxygen, but Whyte has found microbes living there, surviving off methane that bubbles to the surface.

Images from NASA's Mars Orbiter have shown new gullies forming on Mars. No one knows why, but the presence of springs like Lost Hammer could explain it. If microbes can live in the Lost Hammer spring, maybe they can live on Mars. And if there's life on Mars—no matter how small, or primitive—chances seem much better that, one day, we'll find life on other planets, too.

Mars has been called a spacecraft graveyard; it's littered with the orbiters, landers and unmanned rovers we've sent over the years—we've never been able to bring any of them home. How we'd bring our astronauts home is another question. The technology to launch a rocket off a foreign planet, with its own gravity and atmosphere, doesn't yet exist. So maybe, some say, we should just leave them there.

A one-way mission to Mars would cut costs significantly, since we wouldn't need to send fuel for a return trip, says Paul Davies, a cosmologist at Arizona State

University. And the dangers of space travel, including exposure to radiation and the risks associated with takeoff and landing, would be reduced. "People conclude this is a suicide mission," but it's not, he says. On a foreign planet, life expectancy would be lower, he admits, but sending older colonizers (say, those in their fifties) might mean shaving only a few years off their lives. Once on Mars, any scientist would be "the proverbial kid in the candy store," Davies says. "The science would be stupendous. I'm sure plenty of volunteers would be willing." In her book, Roach cites several who say they'd happily do it. She quotes Valentina Tereshkova, the first woman in space: "I am ready to fly without coming back."

The first trip, as envisioned by Davies, would consist of roughly four astronauts. With refuelling missions sent every few years—and perhaps another group of explorers to join them, a decade or so later—they could "set the foundations for a permanent colony," he says. Eventually, they'd have to wean themselves off supplies from Earth, since restocking missions are expensive, so they'd learn to derive what they need from the planet itself; oxygen, Lee says, could be made by breaking down carbon dioxide in the Martian atmosphere. One day, they might even grow their own food in inflatable, radiation-proof greenhouses.

Mike Dixon at the University of Guelph has spent nearly two decades working out how to farm crops in space. In hypobaric chambers at Guelph's Controlled Environment Systems Research Facility, which he directs, his team is tinkering with pressure, lighting and other factors to determine what plants, and the microbes and pollinators that support them, can withstand. (One day in late July, the chambers were full of soybeans.) So far, "we can grow plants at one-tenth of the Earth's atmospheric pressure, with one-third of Earth's oxygen," says Dixon, who works with the CSA and NASA. "They're still perfectly edible." Up on Devon Island, at the Arthur Clarke Mars Greenhouse, his team experiments with germinating and growing crops by remote control before the crew arrives on site, as might be done on a Mars base in the future, he says.

Farming crops in space would have all sorts of benefits, Dixon says, from filtering air to providing the calming effects that come from tending to plants. He hopes to eventually sprout a plant on the moon as proof it can be done.

Before humans ever set foot on Mars, robots will continue to pave the way. Several exploratory missions are planned, with Canadian components. Set to launch next year, NASA's Curiosity—the size of a minivan, it's the largest Mars rover yet—features a Canadian-made tool to measure rock composition. A Canadian team is developing an instrument called MATMOS, to be used on the 2016 ExoMars Trace Gas Orbiter mission (a joint NASA-ESA project) to measure methane in the Martian atmosphere. In 2018, space agencies hope to land a rover on Mars to collect samples. These will be picked up on a future mission and shot back to Earth, which has never yet been done. If successful, says the ESA's Jorge Vago, many of the technologies needed for a manned mission—precision landing, takeoff from Mars and the return to Earth—will be demonstrated on a small scale, for the first time.

When we finally do take off for Mars, it will be our greatest adventure yet. "Ask anyone what happened in the 15th century and they'll say, Columbus discovered the new world," says American planetary scientist Alan Stern. "In the same way, the first mission to Mars will go down in history, I think, in a way people in the 26th century will remember."

He and others permit themselves to dream of a future—maybe not so far away—where Mars is within reach. One day, Zurek muses, maybe we could have "little museums at the Viking and Mir landing sites."

Why We Shouldn't Go to Mars*

By Gregg Easterbrook
Time, January 26, 2004

"Two centuries ago, Meriwether Lewis and William Clark left St. Louis to explore the new lands acquired in the Louisiana Purchase," George W. Bush said, announcing his desire for a program to send men and women to Mars. "They made that journey in the spirit of discovery . . . America has ventured forth into space for the same reasons."

Yet there are vital differences between Lewis and Clark's expedition and a Mars mission. First, Lewis and Clark were headed to a place amenable to life; hundreds of thousands of people were already living there. Second, Lewis and Clark were certain to discover places and things of immediate value to the new nation. Third, the Lewis and Clark venture cost next to nothing by today's standards. In 1989 NASA estimated that a people-to-Mars program would cost $400 billion, which inflates to $600 billion today. The Hoover Dam cost $700 million in today's money, meaning that sending people to Mars might cost as much as building about 800 new Hoover Dams. A Mars mission may be the single most expensive nonwartime undertaking in U.S. history.

The thought of travel to Mars is exhilarating. Surely men and women will someday walk upon that planet, and surely they will make wondrous discoveries about geology and the history of the solar system, perhaps even about the very origin of life. Many times I have stared up at Mars in the evening sky—in the mountains, away from cities, you can almost see the red tint—and wondered what is there, or was there.

But the fact that a destination is tantalizing does not mean the journey makes sense, even considering the human calling to explore. And Mars as a destination for people makes absolutely no sense with current technology.

Present systems for getting from Earth's surface to low-Earth orbit are so fantastically expensive that merely launching the 1,000 tons or so of spacecraft and equipment a Mars mission would require could be accomplished only by cutting health-care benefits, education spending or other important programs—or by raising taxes. Absent some remarkable discovery, astronauts, geologists and biologists once on Mars could do little more than analyze rocks and feel awestruck beholding the sky of another world. Yet rocks can be analyzed by automated probes without risk to human life, and at a tiny fraction of the cost of sending people.

It is interesting to note that when President Bush unveiled his proposal, he listed these recent major achievements of space exploration: pictures of the rings of Saturn and the outer planets, evidence of water on Mars and the moons of Jupiter, discovery of more than 100 planets outside our solar system and study of the soil of Mars. All these accomplishments came from automated probes or automated space telescopes. Bush's proposal, which calls for "reprogramming" some of NASA's present budget into the Mars effort, might actually lead to a reduction in such unmanned science—the one aspect of space exploration that's working really well.

Rather than spend hundreds of billions of dollars to hurl tons toward Mars using current technology, why not take a decade—or two decades, or however much time is required—researching new launch systems and advanced propulsion? If new launch systems could put weight into orbit affordably, and if advanced propulsion could speed up that long, slow transit to Mars, then the dream of stepping onto the Red Planet might become reality. Mars will still be there when the technology is ready.

Space-exploration proponents deride as lack of vision the mention of technical barriers or the insistence that needs on Earth come first. Not so. The former is rationality, the latter the setting of priorities. If Mars proponents want to raise $600 billion privately and stage their own expedition, more power to them; many of the great expeditions of the past were privately mounted. If Mars proponents expect taxpayers to foot their bill, then they must make their case against the many other competing needs for money. And against the needs for health care, education, poverty reduction, reinforcement of the military and reduction of the federal deficit, the case for vast expenditures to go to Mars using current technology is very weak.

The drive to explore is part of what makes us human, and exploration of the past has led to unexpected glories. Dreams must be tempered by realism, however. For the moment, going to Mars is hopelessly unrealistic.

Mars in 39 Days*

By Sam Howe Verhovek
Popular Science, November 2010

You might expect to find our brightest hope for sending astronauts to other planets in Houston, at NASA's Johnson Space Center, inside a high-security multibillion-dollar facility. But it's actually a few miles down the street, in a large warehouse behind a strip mall. This bland and uninviting building is the private aerospace start-up Ad Astra Rocket Company, and inside, founder Franklin Chang Díaz is building a rocket engine that's faster and more powerful than anything NASA has ever flown before. Speed, Chang Díaz believes, is the key to getting to Mars alive.

In fact, he tells me as we peer into a three-story test chamber, his engine will one day travel not just to the Red Planet, but to Jupiter and beyond.

I look skeptical, and Chang Díaz smiles politely. He's used to this reaction. He has been developing the concept of a plasma rocket since 1973, when he became a doctoral student at the Massachusetts Institute of Technology. His idea was this: Rocket fuel is a heavy and inefficient propellant. So instead he imagined building a spaceship engine that uses nuclear reactors to heat plasma to two million degrees. Magnetic fields would eject the hot gas out of the back of the engine. His calculations showed that a spaceship using such an engine could reach 123,000 miles per hour—New York to Los Angeles in about a minute.

Chang Díaz has spent nearly his entire career laboring to convince anyone who would listen that his idea will work, but that career has also taken several turns in the process. One day in 1980, he was pitching the unlimited potential of plasma rockets to yet another MIT professor. The professor listened patiently. "It sounds like borderline science fiction, I know," Chang Díaz was saying. Then the telephone rang. The professor held up a finger. "Why, yes, he's right here," the surprised engineer said into the receiver, then handed it over. "Franklin, it's for you." NASA was on the line. The standout student from Costa Rica had been selected to become an astronaut, the first naturalized American ever chosen for NASA's most elite corps. "I was so excited, I was practically dancing," Chang Díaz recalls. "I almost accidentally strangled my professor with the telephone cord."

All astronauts have big dreams, but Franklin Chang Díaz's dreams are huge. As a college student, as a 25-year astronaut and as an entrepreneur, his single animating intention has always been to build—and fly—a rocketship to Mars. "Of course I wanted to be an astronaut, and of course I want to be able to fly in this," he says of his plasma-thrust rocket. "I mean, I just can't imagine *not* flying in a rocket I would build." And now he's close. In four years Chang Díaz will deploy his technology for the first time in space, when his company, aided by up to $100 million in private funding, plans to test a small rocket on the International Space Station. If this rocket, most commonly known by its loose acronym, Vasimr, for Variable Specific Impulse Magnetoplasma Rocket, proves itself worthy, he has an aggressive timetable for constructing increasingly bigger plasma-thrust space vehicles.

Chang Díaz describes his dreams in relatively practical terms. He doesn't intend to go straight to Mars. First he will develop rockets that perform the more quotidian aspects of space maintenance needed by private companies and by the government: fixing, repositioning, or reboosting wayward satellites; clearing out the ever-growing whirl of "space junk" up there; fetching the stuff that can be salvaged. "Absolutely, fine, I'm not too proud to say it. We're basically running a trucking business here," he says. "We'll be sort of a Triple-A tow truck in space. We're happy to be a local garbage collector in space. That's a reliable, sustainable, affordable business, and that's how you grow."

Eventually, though, Chang Díaz intends to build more than an extraterrestrial trucking business, and his ambitions happen to coincide with Barack Obama's call for a privatized space industry that supports exploration well beyond the moon. "We'll start by sending astronauts to an asteroid for the first time in history," Obama said in a major NASA-related address earlier this year at Kennedy Space Center. "By the mid-2030s, I believe we can send humans to orbit Mars and return them safely to Earth."

Such a belief may seem overly ambitious, but the goals of aviation have always seemed that way. In October 1903, for instance, astronomer Simon Newcomb, the founding president of the American Astronomical Society, spelled out a series of reasons why the concept of powered flight was dubious. "May not our mechanicians," he asked, "be ultimately forced to admit that aerial flight is one of the great class of problems with which man can never cope, and give up all attempts to grapple with it?" Less than two months later, the Wright brothers flew at Kitty Hawk. And in the 1920s a young man named Frank Whittle was coming up with drawings for a theoretical engine very different from the propeller-driven kind, one that might scoop in air through turbines and fire it through a series of "jet" nozzles. "Very interesting, Whittle, my boy," said one of his professors of aeronautical engineering at the University of Cambridge. "But it will never work."

THE CHASE

Chang Díaz decided he wanted to be an astronaut at the age of seven, when his mother explained the U.S.-Soviet space race to him. He went out that night to stare at the stars and look for *Sputnik*. But he soon realized that he had a problem. He happened to be a citizen of Costa Rica, and Costa Rica had no space program. When he was a teenager, he sent a letter to NASA asking how to become an astronaut. He got a letter back saying that to fly with NASA, he must be an American citizen. Boldness runs in his family—his Chinese grandfather fought against the Qing dynasty and fled to Costa Rica during a crackdown on the Nationalist movement. So Chang Díaz was not about to let a thing like citizenship deter him.

After graduating high school, intent on joining NASA, Chang Díaz went to live with relatives in Connecticut. Despite his limited English, he won a University of Connecticut scholarship, one reserved for American citizens. Somebody somehow thought Costa Rica was Puerto Rico, he recalls with a laugh, and after the mistake was pointed out, he was told that the scholarship was being withdrawn. He appealed to university administrators, who agreed to take up his cause with the state, and part of the scholarship was restored, enough to make it possible for him to attend college.

His work was so outstanding there that he was accepted into MIT's doctoral program for nuclear engineering. Then he applied to be an astronaut. NASA turned him down. (Chang Díaz says it was probably because his application for U.S. citizenship had yet to go through.) After he became naturalized—he now holds dual American-Costa Rican citizenship—he tried again, as one of nearly 4,000 applicants for 19 open positions. His plasma-physics doctorate, his singular focus on spaceship engines, his superb physical condition and his obvious drive combined to make him one of NASA's select.

Being an astronaut, Chang Díaz says now, helped him to focus even more on his own vision, and it left him with a much stronger belief that speed was of the essence to get to Mars and beyond. "Believe me," he says, "nobody will want to sit around in a spacecraft for six to eight months if you know you can get there faster."

In many ways, Chang Díaz sees long-range space travel as the ultimate solution to the ultimate problem. The human race, he argues, will one day inevitably conclude that it has to live elsewhere in order to survive. It is also very possible, Chang Díaz concludes, that as resources dry up on Earth, other, potentially more profitable ones may be out there in the cosmos—something vastly more useful for batteries than lithium, perhaps, or for conductivity than copper.

"Whether we'll find gold or riches or something we can't even imagine," he says, "we'll never know until we arrive there to find out. I really don't even see Mars as the end point; I see it as more of a waypoint. We'll open up the entire solar system. Someday we'll find life out there, and everything will change."

Chang Díaz compares space exploration as it has been practiced thus far to exploration as it was practiced in the early years of the American frontier, when Lewis and Clark's government-backed expedition brought back a trove of knowledge about the American West. The next phase, he believes, will be much more like the era of growth in the mid-1800s, when private railroads and mining outfits, helped along by land grants and other government aid, opened the West to epic expansion and settlement.

But, he says, space exploration isn't simply a question of national achievement anymore. "We no longer have a big confrontation between the U.S. and the Soviets," he says. "It's totally different now. We all need each other to make this work." He hopes U.S. citizens will be involved, but even that is by no means certain. "Countries like Brazil, India, China, some in Europe—there's a lot of the same chemistry that the U.S. was feeling, say, 50 or 100 years ago. It's a new club of developers."

To hear Chang Díaz describe it, we're on the verge of a shift from a nationalistic march toward dominance to a much more open and improvisational approach to innovation. If that's true, it's worth noting that he emerged from 25 years at NASA with his sense of improvisation intact.

In 2000, as part of a joint American-Russian survival-training program for crews visiting the ISS, Chang Díaz flew 60 miles west of Moscow for a round of practice drills. The instructors informed him and his two fellow crew members—Kalpana Chawla, killed in 2003 aboard the space shuttle *Columbia*, and a Russian cosmonaut that Chang Díaz describes as being physically much bigger than both of them—that they would need to simulate a mishap that had befallen an earlier mission in which the descent capsule had landed in the midst of a blizzard, and it took rescuers 48 hours to reach it. From now on, the Russians told him, all *Soyuz* crew must train for those conditions.

That's a bit much, Chang Díaz remembers thinking. "Why wouldn't I just delay the reentry by a moment and land in Fiji, or the Indian Ocean, or somewhere else warm and pleasant?" But that was the assignment, and so the trio sat in full flight gear inside a half-buried capsule as engineers heated up the exterior to reentry temperatures with a blowtorch. "It was a sauna," he says.

The crew then had to remove their suits and don survival gear. The capsule was so tight that the process took the better part of a day, with each astronaut taking turns being helped out of one suit and into the other. Finally, they emerged into blinding snow and looked at their manual, which told them to build a shelter. Chang Díaz rolls his eyes at the memory.

Turning to his freezing teammates, he told them to gather up the capsule's parachute silk. His father-in-law, during hunting trips in Montana, had taught him how to build a winter teepee. With the trunks of nearby birch trees and the silk, he improvised. Half an hour later, he had a fire burning inside, and in short order the three were sitting in their socks, dry and comfortable.

Soon a face poked through the flap. The Russians, watching from a ridge through binoculars, wanted to know what was going on. A teepee was not in any manual.

Was everyone all right? The crew smiled. We're fine, Chang Díaz told them. You look so cold. Come on in. So the observers joined them, sat down, doffed their jackets, and sipped tea.

THE CHALLENGE

Chang Díaz, of all people, knows how hard it is to return safely to Earth. His career was bookended by death—death that could have been his own but for the routine tweaks of NASA's scheduling log. The agency decided to pull him off his first scheduled mission on the space shuttle *Challenger* in 1986 and put him on the mission just before instead. Sixteen days later, newly returned to Earth, he watched his close friends and colleagues perish when *Challenger* exploded 73 seconds after takeoff. He went on to fly a total of seven missions between 1986 and 2002—he's tied for the all-time record among astronauts—and logged 1,601 hours beyond Earth's atmosphere. Then, a few months after his final mission, *Columbia* broke apart during reentry, killing all seven people aboard.

Chang Díaz's invention will do little to reduce the dangers of liftoff. Plasma engines depend on the vacuum of space and still require "venerable chemical rockets," as Chang Díaz calls them, to reach Earth orbit. But outer space is where his work stands to vastly improve the safety of a crew. As he points out, a lot can go wrong en route to another planet. The limitation of space travel with a conventional rocket is that the rocket must use its entire fuel supply at once in a single, controlled explosion to reach Earth orbit. It then coasts along at a mostly uniform speed until it enters Mars's gravity. NASA estimates that such a trip would take about seven months. During that time, Chang Díaz explains, there is no abort procedure. The ship cannot change course. If an accident occurs, Earth would be watching, in a 10-minute communications delay, the slow death of the crew. "Chemical rockets are not going to get us to Mars," he says flatly. "It's just too long a trip."

A plasma rocket like Vasimr, on the other hand, sustains propulsion over the entire journey. It accelerates gradually, reaching a maximum speed of 34 miles per second over 23 days. That's at least four times as fast as any chemical rocket could travel, shaving at least six months off a trip to Mars and minimizing the risk of mechanical dangers, exposure to solar radiation (Chang Díaz's design shields the crew behind hydrogen tanks), bone loss, muscle atrophy or any of a thousand other liabilities along the way. And because propulsion is available throughout the trip, the ship could change course at any time.

But human spaceflight programs are currently built around old-fashioned rocketry. NASA has invested mostly in propulsion systems powered by chemical fuel, and for sensible reasons. Chang Díaz's rocket presents many challenges. For one thing, a Vasimr-powered Mars craft would need several nuclear reactors on board to generate the large amount of electricity required to heat the plasma. NASA set to work on a nuclear reactor for space travel in 2003 but scrapped the project after only two years—the risk of radiation from an explosion or crash was likely too great—and redirected its resources to more conventional propulsion programs. Plus, no one has yet determined how to make certain that plasma gas can be safely channeled through a magnetic field. Or just how the human body might respond to traveling at speeds of up to 34 miles per second. "The reality is, rockets don't always work," says Elon Musk, the driving force behind the rocket company SpaceX, one of the key players in the emerging private space industry. For Musk, who struggled for years to get his *Falcon 1* rocket into orbit, the stakes seem particularly high in the case of rockets carrying nuclear material. "If something goes wrong, you have radioactive debris falling to Earth—you have a disaster," he says.

It's true that conventional rockets would be required to put a Mars-bound plasma ship into orbit, but Chang Díaz disputes the notion that launching Vasimr would pose extra risks. The reactors would remain inactivate until the ship was out of the danger zone for spreading radiation back to Earth, he notes. And NASA has already successfully launched several nuclear-electric probes. Nothing is impossible. "We can do this safely," he says. "Our understanding is evolving all the time, but we know that in order to go far, we have to go fast. That's what Vasimr is all about."

SAM HOWE VERHOVEK, *a former national correspondent for the* New York Times *and the* Los Angeles Times, *is the author of* Jet Age: The Comet, the 707, and the Race to Shrink the World.

NASA's Plans for Manned Expedition to Mars May Include Solar-Powered Spacecraft*

By Brian Palmer
The Washington Post, January 24, 2011

Humankind is looking deeper and deeper into the cosmos and seeing incredible things that were totally unknown just a few years ago. In the past three decades, astronomers have described the nature of gamma ray bursts, proved that the expansion of the universe is accelerating, and helped land rovers on Mars. And yet, while our telescopes can see Earthlike planets more than 20 light-years into the intergalactic abyss, we humans can't seem to get past the moon, a sterile rock orbiting just 238,000 miles from Houston. What's taking so long?

The first problem is fuel. The distance between Earth and Mars changes, since our orbits are elliptical, but the minimum distance is 34 million miles. Fueling a spacecraft over that distance poses a challenge, because as you add fuel, you add weight. As you add weight, you have to add more fuel to propel that weight. At certain distances, the cycle becomes totally unmanageable: Traveling on conventional fuel to Proxima Centauri, the next closet star after our sun, for example, would require a tank larger than the visible universe.

Mars isn't nearly that far, but it's far enough that NASA will have to get creative with fuel management. One strategy NASA is considering involves sending pieces of the Mars Transport Vehicle into space, in a series of seven launches, then assembling them in low-Earth orbit. That way, NASA won't have to find a single heavy launch vehicle capable of hauling all those fuel tanks, plus the fuel needed to get to Mars and back, out of the Earth's powerful gravitational field.

Another benefit of in-space assembly is that engineers wouldn't have to worry about aerodynamics, since air resistance isn't a concern in the vacuum of space. Instead of boasting a sleek, Apollo-like design, the Mars vehicle might look more like a modular house.

NASA scientists are also pondering whether a nuclear reactor or a solar-electric engine would be better than the chemical propellants that took us to the moon. The fuel-free electric option is particularly innovative. While scientists say it would provide less acceleration, it could eventually achieve the same top speeds.

"Think of it as a stock portfolio," says NASA's Bret G. Drake, one of the scientists tasked with constructing a vision for putting humans on Mars. "It builds up slowly over time."

The journey to Mars would probably last seven or eight months. To save on cargo weight, the astronauts would rely on dehydrated snacks. The shuttle might also carry a few tabletop gardening devices to provide fresh greens as a periodic treat.

Scientists say landing on Mars may require some fuel and skillful piloting. When astronauts reenter Earth's atmosphere, they use air resistance to decelerate the craft, flying in a complicated pattern to maximize that braking effect before touching down. Mars has an atmosphere, but it's much thinner than Earth's, so the pilots would likely have to use some fuel to decelerate.

We would have a live communications link to astronauts stepping onto the Martian surface, but it would take between eight to 20 minutes for their historic first words to reach Earth.

For those first words, Bill Nye, executive director of the Planetary Society, favors "For the joy of discovery." His selection may merit special consideration, since he's partially responsible for the first written message on Mars. Nye pushed to have the Mars rovers carry the inscription "To those who visit here, we wish a safe journey and the joy of discovery."

Mars would be able to supply astronauts with some useful resources, according to scientists. For example, NASA would send technology enabling them to extract breathable oxygen from the carbon-dioxide-heavy Mars atmosphere. They can also combine carbon dioxide with hydrogen to make water. (NASA assumes that any water the astronauts find is unlikely to be drinkable.)

The astronauts would spend about 18 months on Mars, living in a modular home on the surface of the planet. The timeframe is fairly rigid, because they would have to return when Earth and Mars are close together in their orbits.

Finding the right kind of vehicle for tooling around the Martian surface is key for NASA engineers. For safety reasons, the astronauts would want to land on flat surfaces. Unfortunately, according to Drake, "flat means boring" in exploration terms. "We're looking for areas of geologic diversity."

The astronauts would commute to the more interesting mountain peaks and lava flows in wheeled rovers. Current thinking suggests four two-person machines capable of dragging heavy drilling machinery and covering about 60 miles on a single nuclear-supplied charge.

You might be wondering why it's necessary to send a crew of humans to drill holes in the ground and dig up dirt samples when we already have sent robotic rovers to scour the surface.

Robots, while important, have their limits. For one thing, they're not nearly as resourceful as a human scientists. The rovers extensively survey the topography before they decide to make a move, but they still sometimes get stuck and occasionally can't free themselves. (NASA's Spirit rover is currently mired in a Martian sand trap, although it did outlive its original mission by six years.) A couple of arms and a winch can go a long way on Mars.

In addition, humans are more efficient explorers. "Researchers estimate that what our very best robots do in a week of time, a human can do in about a minute," says Nye. Humans are adaptable and are able to apply their knowledge to new discoveries, while a robot relies on instructions from Earth.

In Nye's view, if a Mars rover reveals a good reason to go to Mars, such as potentially life-sustaining liquid or slushy water, we'll have a scientific obligation to send a human explorer. If you don't wonder whether we are alone in the universe and you're not intrigued by the possibility that terrestrial life could even have come from Mars itself, Nye asks, "why get up in the morning?"

Think you'd like to be the first man or woman to set foot on the Red Planet? If you're out of college, you're probably too late. NASA doesn't anticipate launching a manned mission to Mars until the mid-2020s at the earliest, and the average astronaut candidate historically has been just 34 years old. Better hope NASA develops a time machine to go along with the Mars shuttle.

Making Mars the New Earth*

By Robert Kunzig
National Geographic, February 2010

Could we "terraform" Mars—that is, transform its frozen, thin-aired surface into something more friendly and Earthlike? Should we? The first question has a clear answer: Yes, we probably could. Spacecraft, including the ones now exploring Mars, have found evidence that it was warm in its youth, with rivers draining into vast seas. And right here on Earth, we've learned how to warm a planet: just add greenhouse gases to its atmosphere. Much of the carbon dioxide that once warmed Mars is probably still there, in frozen dirt and polar ice caps, and so is the water. All the planet needs to recapture its salad days is a gardener with a big budget.

Most of the work in terraforming, says NASA planetary scientist Chris McKay, would be done by life itself. "You don't build Mars," McKay says. "You just warm it up and throw some seeds." Perfluorocarbons, potent greenhouse gases, could be synthesized from elements in Martian dirt and air and blown into the atmosphere; by warming the planet, they would release the frozen CO_2, which would amplify the warming and boost atmospheric pressure to the point where liquid water could flow. Meanwhile, says botanist James Graham of the University of Wisconsin, human colonists could seed the red rock with a succession of ecosystems—first bacteria and lichens, which survive in Antarctica, later mosses, and after a millennium or so, redwoods. Coaxing breathable oxygen levels out of those forests, though, could take many millennia.

Enthusiasts such as Robert Zubrin, president of the Mars Society, still dream of Martian cities; Zubrin, an engineer, believes civilization cannot thrive without limitless expansion. Only research outposts seem plausible to McKay. "We're going to live on Mars the way we live in Antarctica," he says. "There are no elementary schools in Antarctica." But he thinks the lessons learned in terraforming Mars—a horrifying prospect to some—would help us manage our limited Earth better.

There is time to debate the point; Mars is in no immediate danger. A White House-appointed panel recently recommended going to the moon or an asteroid first—and pointed out the space agency lacks the budget to go anywhere. It didn't estimate the cost of gardening a dead planet.

YEAR ZERO

1. **THE THOUSAND-YEAR PROJECT** might begin with a series of 18-month survey missions. Each crew making the six-month journey from Earth to Mars would add a small habitation module to the base.

100 YEARS

2. **AN ATMOSPHERE** could be made by releasing carbon dioxide now frozen in dirt and polar ice caps. Factories spewing potent greenhouse gases, and maybe space mirrors focusing sunlight on ice, could start the thaw.

200 YEARS

3. **RAIN** would fall and water would flow once enough CO_2 had been released to raise the atmospheric pressure and warm the planet above freezing. Microbes, algae, and lichens could start taming the desert rock.

600 YEARS

4. **FLOWERING PLANTS** could be introduced after the microbes had created organic soil and added some oxygen to the atmosphere. Boreal and perhaps even temperate forests might ultimately take root.

900 YEARS

5. **ENERGY FOR CITIES**, if a purpose and a desire for them emerged, might come initially from nuclear power and wind turbines. Fusion reactors, if they could be built, might be the best bet in the long run.

1,000 YEARS

6. **MARTIANS** would go out with scuba gear—oxygen would remain low for millennia. Over geologic time, before Earth itself becomes uninhabitable, Mars would lose its new atmosphere and freeze again.

Appendix

Courtesy of NASA/JPL-Caltech/NRL/GSFC

This image was taken by the SECCHI Extreme UltraViolet Imager (EUVI) mounted on the *STEREO-B* spacecraft. *STEREO-B* is located behind the Earth, and follows the Earth in orbit around the sun.

Courtesy of NASA/Cory Huston

A canister, carrying the Alpha Magnetic Spectrometer-2 (AMS) and Express Logistics Carrier-3 for space shuttle *Endeavour's* STS-134 mission, moves past the Vehicle Assembly Building on its journey from the Canister Rotation Facility to Launch Pad 39A at NASA's Kennedy Space Center in Florida.

National Space Policy of the United States of America

June 28, 2010

INTRODUCTION

> "More than by any other imaginative concept, the mind of man is aroused by the thought of exploring the mysteries of outer space. Through such exploration, man hopes to broaden his horizons, add to his knowledge, improve his way of living on earth."
>
> —President Dwight Eisenhower, June 20, 1958

> "Fifty years after the creation of NASA, our goal is no longer just a destination to reach. Our goal is the capacity for people to work and learn and operate and live safely beyond the Earth for extended periods of time, ultimately in ways that are more sustainable and even indefinite. And in fulfilling this task, we will not only extend humanity's reach in space—we will strengthen America's leadership here on Earth."
>
> —President Barack Obama, April 15, 2010

The space age began as a race for security and prestige between two superpowers. The opportunities were boundless, and the decades that followed have seen a radical transformation in the way we live our daily lives, in large part due to our use of space. Space systems have taken us to other celestial bodies and extended humankind's horizons back in time to the very first moments of the universe and out to the galaxies at its far reaches. Satellites contribute to increased transparency and stability among nations and provide a vital communications path for avoiding potential conflicts. Space systems increase our knowledge in many scientific fields, and life on Earth is far better as a result.

The utilization of space has created new markets; helped save lives by warning us of natural disasters, expediting search and rescue operations, and making recovery efforts faster and more effective; made agriculture and natural resource management more efficient and sustainable; expanded our frontiers; and provided global access to advanced medicine, weather forecasting, geospatial information, financial operations, broadband and other communications, and scores of other activities worldwide. Space systems allow people and governments around the world to see with clarity, communicate with certainty, navigate with accuracy, and operate with assurance.

The legacy of success in space and its transformation also presents new challenges. When the space age began, the opportunities to use space were limited to only a few nations, and there were limited consequences for irresponsible or unintentional behavior. Now, we find ourselves in a world where the benefits of space permeate almost every facet of our lives. The growth and evolution of the global economy has ushered in an ever-increasing number of nations and organizations using space. The now-ubiquitous and interconnected nature of space capabilities and the world's growing dependence on them mean that irresponsible acts in space can have damaging consequences for all of us. For example, decades of space activity have littered Earth's orbit with debris; and as the world's space-faring nations continue to increase activities in space, the chance for a collision increases correspondingly.

As the leading space-faring nation, the United States is committed to addressing these challenges. But this cannot be the responsibility of the United States alone. All nations have the right to use and explore space, but with this right also comes responsibility. The United States, therefore, calls on all nations to work together to adopt approaches for responsible activity in space to preserve this right for the benefit of future generations.

From the outset of humanity's ascent into space, this Nation declared its commitment to enhance the welfare of humankind by cooperating with others to maintain the freedom of space.

The United States hereby renews its pledge of cooperation in the belief that with strengthened international collaboration and reinvigorated U.S.leadership, all nations and peoples—space-faring and space-benefiting—will find their horizons broadened, their knowledge enhanced, and their lives greatly improved.

PRINCIPLES

In this spirit of cooperation, the United States will adhere to, and proposes that other nations recognize and adhere to, the following principles:

- It is the shared interest of all nations to act responsibly in space to help prevent mishaps, misperceptions, and mistrust. The United States considers the sustainability, stability, and free access to, and use of, space vital to its national interests. Space operations should be conducted in ways that emphasize openness and transparency to improve public awareness of the activities of government, and enable others to share in the benefits provided by the use of space.
- A robust and competitive commercial space sector is vital to continued progress in space. The United States is committed to encouraging and facilitating the growth of a U.S. commercial space sector that supports U.S.

needs, is globally competitive, and advances U.S. leadership in the generation of new markets and innovation-driven entrepreneurship.

- All nations have the right to explore and use space for peaceful purposes, and for the benefit of all humanity, in accordance with international law. Consistent with this principle, "peaceful purposes" allows for space to be used for national and homeland security activities.
- As established in international law, there shall be no national claims of sovereignty over outer space or any celestial bodies. The United States considers the space systems of all nations to have the rights of passage through, and conduct of operations in, space without interference. Purposeful interference with space systems, including supporting infrastructure, will be considered an infringement of a nation's rights.
- The United States will employ a variety of measures to help assure the use of space for all responsible parties, and, consistent with the inherent right of self-defense, deter others from interference and attack, defend our space systems and contribute to the defense of allied space systems, and, if deterrence fails, defeat efforts to attack them.

GOALS

Consistent with these principles, the United States will pursue the following goals in its national space programs:

- **Energize competitive domestic industries** to participate in global markets and advance the development of: satellite manufacturing; satellite-based services; space launch; terrestrial applications; and increased entrepreneurship.
- **Expand international cooperation on** mutually beneficial space activities to: broaden and extend the benefits of space; further the peaceful use of space; and enhance collection and partnership in sharing of space-derived information.
- **Strengthen stability in space** through: domestic and international measures to promote safe and responsible operations in space; improved information collection and sharing for space object collision avoidance; protection of critical space systems and supporting infrastructures, with special attention to the critical interdependence of space and information systems; and strengthening measures to mitigate orbital debris.
- **Increase assurance and resilience of mission-essential functions enabled** by commercial, civil, scientific, and national security spacecraft and supporting infrastructure against disruption, degradation, and destruction, whether from environmental, mechanical, electronic, or hostile causes.
- **Pursue human and robotic initiatives** to develop innovative technologies, foster new industries, strengthen international partnerships, inspire our Nation and the world, increase humanity's understanding of the Earth, en-

hance scientific discovery, and explore our solar system and the universe beyond.

- **Improve space-based Earth and solar observation** capabilities needed to conduct science, forecast terrestrial and near-Earth space weather, monitor climate and global change, manage natural resources, and support disaster response and recovery.

All actions undertaken by departments and agencies in implementing this directive shall be within the overall resource and policy guidance provided by the President; consistent with U.S. law and regulations, treaties and other agreements to which the United States is a party, other applicable international law, U.S. national and homeland security requirements, U.S. foreign policy, and national interests; and in accordance with the Presidential Memorandum on Transparency and Open Government.

INTERSECTOR GUIDELINES

In pursuit of this directive's goals, all departments and agencies shall execute the following guidance:

Foundational Activities and Capabilities

- **Strengthen U.S. Leadership In Space-Related Science, Technology, and Industrial Bases.** Departments and agencies shall: conduct basic and applied research that increases capabilities and decreases costs, where this research is best supported by the government; encourage an innovative and entrepreneurial commercial space sector; and help ensure the availability of space-related industrial capabilities in support of critical government functions.
- **Enhance Capabilities for Assured Access To Space.** United States access to space depends in the first instance on launch capabilities. United States Government payloads shall be launched on vehicles manufactured in the United States unless exempted by the National Security Advisor and the Assistant to the President for Science and Technology and Director of the Office of Science and Technology Policy, consistent with established interagency standards and coordination guidelines. Where applicable to their responsibilities departments and agencies shall:
  - Work jointly to acquire space launch services and hosted payload arrangements that are reliable, responsive to United States Government needs, and cost-effective;
  - Enhance operational efficiency, increase capacity, and reduce launch costs by investing in the modernization of space launch infrastructure; and
  - Develop launch systems and technologies necessary to assure and sustain future reliable and efficient access to space, in cooperation with U.S.

industry, when sufficient U.S. commercial capabilities and services do not exist.

- **Maintain and Enhance Space-based Positioning, Navigation, and Timing Systems.** The United States must maintain its leadership in the service, provision, and use of global navigation satellite systems (GNSS). To this end, the United States shall:
  - Provide continuous worldwide access, for peaceful civil uses, to the Global Positioning System (GPS) and its government-provided augmentations, free of direct user charges;
  - Engage with foreign GNSS providers to encourage compatibility and interoperability, promote transparency in civil service provision, and enable market access for U.S. industry;
  - Operate and maintain the GPS constellation to satisfy civil and national security needs, consistent with published performance standards and interface specifications. Foreign positioning, navigation, and timing (PNT) services may be used to augment and strengthen the resiliency of GPS; and
  - Invest in domestic capabilities and support international activities to detect, mitigate, and increase resiliency to harmful interference to GPS, and identify and implement, as necessary and appropriate, redundant and back-up systems or approaches for critical infrastructure, key resources, and mission-essential functions.
- **Develop and Retain Space Professionals.** The primary goals of space professional development and retention are: achieving mission success in space operations and acquisition; stimulating innovation to improve commercial, civil, and national security space capabilities; and advancing science, exploration, and discovery. Toward these ends, departments and agencies, in cooperation with industry and academia, shall establish standards, seek to create opportunities for the current space workforce, and implement measures to develop, maintain, and retain skilled space professionals, including engineering and scientific personnel and experienced space system developers and operators, in government and commercial workforces. Departments and agencies also shall promote and expand public-private partnerships to foster educational achievement in Science, Technology, Engineering, and Mathematics (STEM) programs, supported by targeted investments in such initiatives.
- Improve Space System Development and Procurement. Departments and agencies shall:
  - Improve timely acquisition and deployment of space systems through enhancements in estimating costs, technological risk and maturity, and industrial base capabilities;
  - Reduce programmatic risk through improved management of requirements and by taking advantage of cost-effective opportunities to test

high-risk components, payloads, and technologies in space or relevant environments;
- Embrace innovation to cultivate and sustain an entrepreneurial U.S.research and development environment; and
- Engage with industrial partners to improve processes and effectively manage the supply chains.

- **Strengthen Interagency Partnerships.** Departments and agencies shall improve their partnerships through cooperation, collaboration, information sharing, and/or alignment of common pursuits. Departments and agencies shall make their capabilities and expertise available to each other to strengthen our ability to achieve national goals, identify desired outcomes, leverage U.S. capabilities, and develop implementation and response strategies.

International Cooperation

Strengthen U.S. Space Leadership. Departments and agencies, in coordination with the Secretary of State, shall:

- Demonstrate U.S. leadership in space-related fora and activities to: reassure allies of U.S. commitments to collective self-defense; identify areas of mutual interest and benefit; and promote U.S. commercial space regulations and encourage interoperability with these regulations;
- Lead in the enhancement of security, stability, and responsible behavior in space;
- Facilitate new market opportunities for U.S. commercial space capabilities and services, including commercially viable terrestrial applications that rely on government-provided space systems;
- Promote the adoption of policies internationally that facilitate full, open, and timely access to government environmental data;
- Promote appropriate cost- and risk-sharing among participating nations in international partnerships; and
- Augment U.S. capabilities by leveraging existing and planned space capabilities of allies and space partners.

Identify Areas for Potential International Cooperation. Departments and agencies shall identify potential areas for international cooperation that may include, but are not limited to: space science; space exploration, including human space flight activities; space nuclear power to support space science and exploration; space transportation; space surveillance for debris monitoring and awareness; missile warning; Earth science and observation; environmental monitoring; satellite communications; GNSS; geospatial information products and services; disaster mitigation and relief; search and rescue; use of space for maritime domain awareness; and long-term preservation of the space environment for human activity and use.

The Secretary of State, after consultation with the heads of appropriate departments and agencies, shall carry out diplomatic and public diplomacy efforts to strengthen understanding of, and support for, U.S. national space policies and programs and to encourage the foreign use of U.S. space capabilities, systems, and services.

Develop Transparency and Confidence-Building Measures. The United States will pursue bilateral and multilateral transparency and confidence-building measures to encourage responsible actions in, and the peaceful use of, space. The United States will consider proposals and concepts for arms control measures if they are equitable, effectively verifiable, and enhance the national security of the United States and its allies.

Preserving the Space Environment and the Responsible Use of Space

Preserve the Space Environment. For the purposes of minimizing debris and preserving the space environment for the responsible, peaceful, and safe use of all users, the United States shall:

- Lead the continued development and adoption of international and industry standards and policies to minimize debris, such as the United Nations Space Debris Mitigation Guidelines;
- Develop, maintain, and use space situational awareness (SSA) information from commercial, civil, and national security sources to detect, identify, and attribute actions in space that are contrary to responsible use and the long-term sustainability of the space environment;
- Continue to follow the United States Government Orbital Debris Mitigation Standard Practices, consistent with mission requirements and cost effectiveness, in the procurement and operation of spacecraft, launch services, and the conduct of tests and experiments in space;
- Pursue research and development of technologies and techniques, through the Administrator of the National Aeronautics and Space Administration (NASA) and the Secretary of Defense, to mitigate and remove on-orbit debris, reduce hazards, and increase understanding of the current and future debris environment; and
- Require the head of the sponsoring department or agency to approve exceptions to the United States Government Orbital Debris Mitigation Standard Practices and notify the Secretary of State.

Foster the Development of Space Collision Warning Measures. The Secretary of Defense, in consultation with the Director of National Intelligence, the Administrator of NASA, and other departments and agencies, may collaborate with industry and foreign nations to: maintain and improve space object databases; pursue common international data standards and data integrity measures; and provide services and disseminate orbital tracking information to commercial and international entities, including predictions of space object conjunction.

Effective Export Policies

Consistent with the U.S. export control review, departments and agencies should seek to enhance the competitiveness of the U.S.space industrial base while also addressing national security needs.

The United States will work to stem the flow of advanced space technology to unauthorized parties. Departments and agencies are responsible for protecting against adverse technology transfer in the conduct of their programs.

The United States Government will consider the issuance of licenses for space-related exports on a case-by-case basis, pursuant to, and in accordance with, the International Traffic in Arms Regulations, the Export Administration Regulations, and other applicable laws, treaties, and regulations. Consistent with the foregoing space-related items that are determined to be generally available in the global marketplace shall be considered favorably with a view that such exports are usually in the national interests of the United States.

Sensitive or advanced spacecraft-related exports may require a government-to-government agreement or other acceptable arrangement.

Space Nuclear Power

The United States shall develop and use space nuclear power systems where such systems safely enable or significantly enhance space exploration or operational capabilities.

Approval by the President or his designee shall be required to launch and use United States Government spacecraft utilizing nuclear power systems either with a potential for criticality or above a minimum threshold of radioactivity, in accordance with the existing interagency review process. To inform this decision, the Secretary of Energy shall conduct a nuclear safety analysis for evaluation by an ad hoc Interagency Nuclear Safety Review Panel that will evaluate the risks associated with launch and in-space operations.

The Secretary of Energy shall:

- Assist the Secretary of Transportation in the licensing of space transportation activities involving spacecraft with nuclear power systems;
- Provide nuclear safety monitoring to ensure that operations in space are consistent with any safety evaluations performed; and
- Maintain the capability and infrastructure to develop and furnish nuclear power systems for use in United States Government space systems.

Radiofrequency Spectrum and Interference Protection

The United States Government shall:

- Seek to protect U.S. global access to, and operation in, the radiofrequency spectrum and related orbital assignments required to support the use of space by the United States Government, its allies, and U.S. commercial users;

- Explicitly address requirements for radiofrequency spectrum and orbital assignments prior to approving acquisition of space capabilities;
- Seek to ensure the necessary national and international regulatory frameworks will remain in place over the lifetime of the system;
- Identify impacts to government space systems prior to reallocating spectrum for commercial, federal, or shared use;
- Enhance capabilities and techniques, in cooperation with civil, commercial, and foreign partners, to identify, locate, and attribute sources of radio frequency interference, and take necessary measures to sustain the radiofrequency environment in which critical U.S. space systems operate; and
- Seek appropriate regulatory approval under U.S. domestic regulations for United States Government earth stations operating with commercially owned satellites, consistent with the regulatory approval granted to analogous commercial earth stations.

Assurance and Resilience of Mission-Essential Functions

The United States shall:

- Assure space-enabled mission-essential functions by developing the techniques, measures, relationships, and capabilities necessary to maintain continuity of services;
  - Such efforts may include enhancing the protection and resilience of selected spacecraft and supporting infrastructure;
- Develop and exercise capabilities and plans for operating in and through a degraded, disrupted, or denied space environment for the purposes of maintaining mission-essential functions; and
- Address mission assurance requirements and space system resilience in the acquisition of future space capabilities and supporting infrastructure.

SECTOR GUIDELINES

United States space activities are conducted in three distinct but interdependent sectors: commercial, civil, and national security.

Commercial Space Guidelines

The term "commercial," for the purposes of this policy, refers to space goods, services, or activities provided by private sector enterprises that bear a reasonable portion of the investment risk and responsibility for the activity, operate in accordance with typical market-based incentives for controlling cost and optimizing return on investment, and have the legal capacity to offer these goods or services to existing or potential nongovernmental customers. To promote a robust domestic commercial space industry, departments and agencies shall:

- Purchase and use commercial space capabilities and services to the maximum practical extent when such capabilities and services are available in the marketplace and meet United States Government requirements;
- Modify commercial space capabilities and services to meet government requirements when existing commercial capabilities and services do not fully meet these requirements and the potential modification represents a more cost-effective and timely acquisition approach for the government;
- Actively explore the use of inventive, nontraditional arrangements for acquiring commercial space goods and services to meet United States Government requirements, including measures such as public-private partnerships, hosting government capabilities on commercial spacecraft, and purchasing scientific or operational data products from commercial satellite operators in support of government missions;
- Develop governmental space systems only when it is in the national interest and there is no suitable, cost-effective U.S. commercial or, as appropriate, foreign commercial service or system that is or will be available;
- Refrain from conducting United States Government space activities that preclude, discourage, or compete with U.S. commercial space activities, unless required by national security or public safety;
- Pursue potential opportunities for transferring routine, operational space functions to the commercial space sector where beneficial and cost-effective, except where the government has legal, security, or safety needs that would preclude commercialization;
- Cultivate increased technological innovation and entrepreneurship in the commercial space sector through the use of incentives such as prizes and competitions;
- Ensure that United States Government space technology and infrastructure are made available for commercial use on a reimbursable, noninterference, and equitable basis to the maximum practical extent;
- Minimize, as much as possible, the regulatory burden for commercial space activities and ensure that the regulatory environment for licensing space activities is timely and responsive;
- Foster fair and open global trade and commerce through the promotion of suitable standards and regulations that have been developed with input from U.S. industry;
- Encourage the purchase and use of U.S. commercial space services and capabilities in international cooperative arrangements; and
- Actively promote the export of U.S. commercially developed and available space goods and services, including those developed by small- and medium-sized enterprises, for use in foreign markets, consistent with U.S. technology transfer and nonproliferation objectives.

The United States Trade Representative (USTR) has the primary responsibility in the Federal Government for international trade agreements to which the United States is a party. USTR, in consultation with other relevant departments and agencies, will lead any efforts relating to the negotiation and implementation of trade disciplines governing trade in goods and services related to space.

Civil Space Guidelines

Space Science, Exploration, and Discovery

The Administrator of NASA shall:

- Set far-reaching exploration milestones. By 2025, begin crewed missions beyond the moon, including sending humans to an asteroid. By the mid-2030s, send humans to orbit Mars and return them safely to Earth;
- Continue the operation of the International Space Station (ISS), in cooperation with its inter-national partners, likely to 2020 or beyond, and expand efforts to: utilize the ISS for scientific, technological, commercial, diplomatic, and educational purposes; support activities requiring the unique attributes of humans in space; serve as a continuous human presence in Earth orbit; and support future objectives in human space exploration;
- Seek partnerships with the private sector to enable safe, reliable, and cost-effective commercial spaceflight capabilities and services for the transport of crew and cargo to and from the ISS;
- Implement a new space technology development and test program, working with industry, academia, and international partners to build, fly, and test several key technologies that can increase the capabilities, decrease the costs, and expand the opportunities for future space activities;
- Conduct research and development in support of next-generation launch systems, including new U.S. rocket engine technologies;
- Maintain a sustained robotic presence in the solar system to: conduct scientific investigations of other planetary bodies; demonstrate new technologies; and scout locations for future human missions;
- Continue a strong program of space science for observations, research, and analysis of our Sun, solar system, and universe to enhance knowledge of the cosmos, further our understanding of fundamental natural and physical sciences, understand the conditions that may support the development of life, and search for planetary bodies and Earth-like planets in orbit around other stars; and
- Pursue capabilities, in cooperation with other departments, agencies, and commercial partners, to detect, track, catalog, and characterize near-Earth objects to reduce the risk of harm to humans from an unexpected impact on our planet and to identify potentially resource-rich planetary objects.

Environmental Earth Observation and Weather

To continue and improve a broad array of programs of space-based observation, research, and analysis of the Earth's land, oceans, and atmosphere:

- The NASA Administrator, in coordination with other appropriate departments and agencies, shall conduct a program to enhance U.S. global climate change research and sustained monitoring capabilities, advance research into and scientific knowledge of the Earth by accelerating the development of new Earth observing satellites, and develop and test capabilities for use by other civil departments and agencies for operational purposes.

- The Secretary of Commerce, through the National Oceanic and Atmospheric Administration (NOAA) Administrator, and in coordination with the NASA Administrator and other appropriate departments and agencies, shall, in support of operational requirements:
  - Transition mature research and development Earth observation satellites to long-term operations;
  - Use international partnerships to help sustain and enhance weather, climate, ocean, and coastal observation from space; and
  - Be responsible for the requirements, funding, acquisition, and operation of civil operational environmental satellites in support of weather forecasting, climate monitoring, ocean and coastal observations, and space weather forecasting.NOAA will primarily utilize NASA as the acquisition agent for operational environmental satellites for these activities and programs.
- The Secretary of Commerce, through the NOAA Administrator, the Secretary of Defense, through the Secretary of the Air Force, and the NASA Administrator shall work together and with their international partners to ensure uninterrupted, operational polar-orbiting environmental satellite observations. The Secretary of Defense shall be responsible for the morning orbit, and the Secretary of Commerce shall be responsible for the afternoon orbit. The departments shall continue to partner in developing and fielding a shared ground system, with the coordinated programs operated by NOAA. Further, the departments shall ensure the continued full sharing of data from all systems.

Land Remote Sensing

The Secretary of the Interior, through the Director of the United States Geological Survey (USGS), shall:

- Conduct research on natural and human-induced changes to Earth's land, land cover, and inland surface waters, and manage a global land surface data national archive and its distribution;
- Determine the operational requirements for collection, processing, archiving, and distribution of land surface data to the United States Government and other users; and
- Be responsible, in coordination with the Secretary of Defense, the Secretary of Homeland Security, and the Director of National Intelligence, for providing remote sensing information related to the environment and disasters that is acquired from national security space systems to other civil government agencies.

In support of these critical needs, the Secretary of the Interior, through the Director of the USGS, and the NASA Administrator shall work together in maintaining a program for operational land remote sensing observations.

The NASA and NOAA Administrators and the Director of the USGS shall:

- Ensure that civil space acquisition processes and capabilities are not unnecessarily duplicated; and
- Continue to develop civil applications and information tools based on data collected by Earth observation satellites. These civil capabilities will be developed, to the greatest extent possible, using known standards and open protocols, and the applications will be made available to the public.

The Secretary of Commerce, through the Administrator of NOAA, shall provide for the regulation and licensing of the operation of commercial sector remote sensing systems.

National Security Space Guidelines

The Secretary of Defense and the Director of National Intelligence, in consultation with other appropriate heads of departments and agencies, shall:

- Develop, acquire, and operate space systems and supporting information systems and networks to support U.S. national security and enable defense and intelligence operations during times of peace, crisis, and conflict;
- Ensure cost-effective survivability of space capabilities, including supporting information systems and networks, commensurate with their planned use, the consequences of lost or degraded capability, the threat, and the availability of other means to perform the mission;
- Reinvigorate U.S. leadership by promoting technology development, improving industrial capacity, and maintaining a robust supplier base necessary to support our most critical national security interests;
- Develop and implement plans, procedures, techniques, and capabilities necessary to assure critical national security space-enabled missions. Options for mission assurance may include rapid restoration of space assets and leveraging allied, foreign, and/or commercial space and nonspace capabilities to help perform the mission;
- Maintain and integrate space surveillance, intelligence, and other information to develop accurate and timely SSA. SSA information shall be used to support national and homeland security, civil space agencies, particularly human space flight activities, and commercial and foreign space operations;
- Improve, develop, and demonstrate, in cooperation with relevant departments and agencies and commercial and foreign entities, the ability to rapidly detect, warn, characterize, and attribute natural and man-made disturbances to space systems of U.S. interest; and
- Develop and apply advanced technologies and capabilities that respond to changes to the threat environment.

The Secretary of Defense shall:

- Be responsible, with support from the Director of National Intelligence, for the development, acquisition, operation, maintenance, and modernization of SSA capabilities;
- Develop capabilities, plans, and options to deter, defend against, and, if necessary, defeat efforts to interfere with or attack U.S. or allied space systems;
- Maintain the capabilities to execute the space support, force enhancement, space control, and force application missions; and
- Provide, as launch agent for both the defense and intelligence sectors, reliable, affordable, and timely space access for national security purposes.

The Director of National Intelligence shall:

- Enhance foundational intelligence collection and single- and all-source intelligence analysis;
- Develop, obtain, and operate space capabilities to support strategic goals, intelligence priorities, and assigned tasks;
- Provide robust, timely, and effective collection, processing, analysis, and dissemination of information on foreign space and supporting information system activities;
- Develop and enhance innovative analytic tools and techniques to use and share information from traditional and nontraditional sources for understanding foreign space-related activities;
- Identify and characterize current and future threats to U.S. space missions for the purposes of enabling effective protection, deterrence, and defense;
- Integrate all-source intelligence of foreign space capabilities and intentions with space surveillance information to produce enhanced intelligence products that support SSA;
- Support national defense and homeland security planning and satisfy operational requirements as a major intelligence mission;
- Support monitoring, compliance, and verification for transparency and confidence-building measures and, if applicable, arms control agreements; and
- Coordinate on any radiofrequency surveys from space conducted by United States Government departments or agencies and review, as appropriate, any radiofrequency surveys from space conducted by licensed private sector operators or by state and local governments.

Bibliography

Courtesy of NASA

This photo shows the Hubble Space Telescope (HST) following grapple.

Courtesy of NASA

The liftoff of *Apollo 13* from Florida's Kennedy Space Center on April 11, 1970.

Books

Belfiore, Michael P. *Life Aboard a Space Station*. Detroit: Lucent Books, Thomson/Gale, 2005.

———. *Rocketeers: How a Visionary Band of Business Leaders, Engineers, and Pilots Is Boldly Privatizing Space*. Washington, D.C.: Smithsonian Books, 2007.

Biddle, Wayne. *Dark Side of the Moon: Wernher von Braun, the Third Reich, and the Space Race*. New York: W. W. Norton, 2009.

Burrows, William E. *The New Ocean: The Story of the First Space Age*. New York: Random House, 1998.

Cernan, Eugene, and Don Davis. *The Last Man on the Moon: Astronaut Eugene Cernan and America's Race in Space*. New York: St. Martin's Press, 1999.

Chaikin, Andrew. *A Man on the Moon: The Voyages of the Apollo Astronauts*. New York: Viking, 1994.

———. *A Passion for Mars: Intrepid Explorers of the Red Planet*. New York: Abrams, 2008.

Collins, Michael. *Carrying the Fire: An Astronaut's Journeys*. New York: Farrar, Straus and Giroux, 1974.

Duggins, Pat. *Final Countdown: NASA and the End of the Space Shuttle Program*. Gainesville, Fla.: University Press of Florida, 2009.

———. *Trailblazing Mars: NASA's Next Giant Leap*. Gainesville, Fla.: University Press of Florida, 2010.

Hansen, James R. *First Man: The Life of Neil A. Armstrong*. New York: Simon & Schuster, 2005.

Kranz, Gene. *Failure Is Not an Option: Mission Control from Mercury to Apollo 13 and Beyond.* New York: Simon & Schuster, 2000.

Light, Michael. *Full Moon.* New York: Alfred A. Knopf, 1999.

Lovell, Jim, and Jeffrey Kluger. *Lost Moon: The Perilous Voyage of Apollo 13.* Boston: Houghton Mifflin, 1994.

McCurdy, Howard E. *Faster, Better, Cheaper: Low-Cost Innovation in the U.S. Space Program.* Baltimore, Md.: John Hopkins University Press, 2001.

Sagan, Carl. *Cosmos.* New York: Random House, 1980.

———. *Pale Blue Dot: A Vision of the Human Future in Space.* New York: Random House, 1994.

Schmidt, Stanley, and Robert Zubrin, eds. *Islands in the Sky: Bold New Ideas for Colonizing Space.* New York: John Wiley & Sons, 1996.

Shepard, Alan, and Deke Slayton. *Moon Shot: The Inside Story of America's Race to the Moon.* Atlanta, Ga.: Turner Pub.; Kansas City, Mo.: Distributed by Andrews and McMeel, 1994.

Shirley, Donna, with Danelle Morton. *Managing Martians.* New York: Broadway Books, 1998.

Slayton, Donald K. "Deke," with Michael Cassutt. *Deke! U.S. Manned Space: From Mercury to the Shuttle.* New York: Forge, 1994.

Wolfe, Tom. *The Right Stuff.* New York: Farrar, Straus and Giroux, 1979.

Zubrin, Robert, with Richard Wagner. *The Case for Mars: The Plan to Settle the Red Planet and Why We Must.* New York: The Free Press, 1996.

Zubrin, Robert. *How to Live on Mars: A Trusty Guidebook to Surviving and Thriving on the Red Planet.* New York: Three Rivers Press, 2008.

Web Sites

Readers seeking additional information on American space exploration and development and related subjects may wish to consult the following Web sites, all of which were operational as of this writing.

Great Images at NASA (GRIN)

http://grin.hq.nasa.gov/

According to its Web site, "GRIN is a collection of over a thousand images of significant historical interest scanned at high-resolution in several sizes. This collection is intended for the media, publishers, and the general public looking for high-quality photographs." The Web site is maintained by NASA.

The Mars Society

http://www.marssociety.org/

According to the society's Web site, "The purpose of the Mars Society is to explore and settle the planet Mars. . . . Starting small, with hitchhiker payloads on government-funded missions, we intend to use the credibility such activity engenders to mobilize larger resources, further enabling private robotic missions and ultimately human exploration and settlement of Mars." In order to advocate for the colonization and possible terraforming of Mars, the society maintains local chapters around the world.

National Aeronautics and Space Administration (NASA)

http://www.nasa.gov/

NASA has been charged with running the U.S. space program since its establishment in July 1958. In addition to overseeing manned and unmanned spaceflight, the agency is also responsible for long-term civilian and military aerospace research. Its Web site provides text, images, and video that present not only an overview of NASA's history but also its current missions.

The National Space Society (NSS)

http://www.nss.org/

According to this organization's Web site, the NSS's vision is, "People living and working in thriving communities beyond the Earth, and the use of the vast resources of space for the dramatic betterment of humanity." NSS publishes *Ad Astra* magazine, maintains a network of local chapters, and advocates for better and more expansive funding of space exploration.

Space.com

http://www.space.com/

Utilizing a multimedia approach, Space.com provides breaking news about outer space, spaceflight, science, and astronomy.

Space Exploration Technologies Corp. (SpaceX)

http://www.spacex.com/

SpaceX is one of the leading private space companies. Its official Web site provides information about its Falcon Rocket family.

Views of the Solar System

http://www.solarviews.com/eng/homepage.html/

Created by Calvin J. Hamilton, Views of the Solar System is one of the most comprehensive space exploration Web sites maintained by a private individual. Hamilton uses photographs, scientific facts, text, graphics, and videos to provide visitors with a planet-by-planet understanding of our solar system.

Additional Periodical Articles with Abstracts

More information about American space exploration and development and related topics can be found in the following articles. Readers interested in additional material may consult the *Readers' Guide to Periodical Literature* and other H.W. Wilson publications.

International Partnerships: The World Is Going Back to the Moon. *Ad Astra* v. 22 p37 Summer 2010.

In January, NASA made a further move to foster international cooperation in lunar science, the writer notes. NASA administrator Charles Bolden and Lunar Science Institute director David Morrison and Assistant Director Greg Schmidt traveled to Israel to sign a joint statement with the Israel Space Agency, which recognized the Israel Network for Lunar Science and Exploration as an affiliate partner with the Ames Research Center. Bolden then went to the Fifth International Ilan Ramon Conference at the Fischer Institute for Strategic Research in Herzliya, held in memory of Israel Air Force Col. Ilan Ramon, who was killed onboard the *Columbia* seven years ago.

Finding Common Ground/Space? Gary Barnhard. *Ad Astra* v. 22 p47 Spring 2010.

The way forward for the United States in the exploration and development of space is at a crucial political and economic crossroads Barnhard reports. There are some who want to press ahead with the current plan or some variation of it, regardless of its viability. There are others who want to change direction altogether, without considering whether such a shift can lead to sustainable programs. Barnhard argues that both of these approaches are flawed and suggests that common ground needs to be found.

A Controversial New Direction for NASA. Jeff Foust. *Ad Astra* v. 22 pp18–21 Spring 2010.

The White House unveiled its latest plans for NASA in the budget proposal for fiscal year 2011, Foust notes. These budget documents, and subsequent statements by NASA and White House officials, represent a dramatic new direction for the agency. The new direction turns away from previous plans to develop systems for returning humans to the moon in favor of developing new technologies and commercial capabilities that can sustain human space exploration over the long run.

America's Space Program for the 21st Century Conference. Lewis Groswald. *Ad Astra* v. 22 pp28–29 Summer 2010.

In one of the more cogent speeches delivered by a president on space policy, Barack Obama told Americans that "for pennies on the dollar," taxpayers get the space shuttle, the International Space Station, and the secrets of the universe, Groswald notes. In reality, it is even less than that, as NASA's budget comes in just under one-quarter of a penny of the taxpayer dollar. Obama was speaking at a conference of around 200 people at John F. Kennedy Space Center in Cape Canaveral, Florida, which came amidst a heated debate over the future of the agency and the country's human space exploration program. According to NASA administrator Major General Charles F. Bolden (U.S. Marines, Ret.), this is an intellectual debate that should have happened four decades ago.

A New Era for Science: Research on the International Space Station. Clifford R. McMurray. *Ad Astra* v. 22 pp24–27 Summer 2010.

From the time the inaugural crew settled in aboard the International Space Station (ISS), McMurray writes, it has functioned as an operational research platform, but until recently, scientific experiment and commercial utilization were secondary to the heavy time demands of the assembly process. Now the decade of construction for the ISS is ending, and the space station will come into its own as a center of technology development and scientific research for all the countries that have partnered to construct it. ISS is currently at its proposed permanent crew level of six astronauts. The amount of crew time devoted to research activity has soared from around 15 hours a week to around 60 hours a week and continues to rise.

A Look at Lunar Exploration for NASA. David Schlom. *Ad Astra* v. 22 pp34–37 Summer 2010.

With the retirement of the Space Shuttle ending an age of low-Earth-orbit expectations, NASA has found a new stride in dramatic scientific discoveries via its existing spacecraft, Schlom reports. From the geysers of Enceladus to the ancient Martian saline sea beds and the rejuvenated Hubble Space Telescope, science is experiencing a golden age of cosmic discovery. The recent finding of water ice and other volatiles in Cabeus crater by the LCROSS mission has prompted even greater interest in future lunar exploration. Science is playing a key role as NASA seeks to reinvent itself in the post-Shuttle age. The Obama administration has revealed its intentions, however, to abandon the Constellation Program and focus U.S. efforts on developing private enterprise access to Earth orbit. Furthermore, officials have directed the space agency to develop the next generation of propulsion systems while maintaining a strong focus on science via robotic exploration of the solar system and monitoring of Earth's climate.

Finding Solutions in Thin Air. Tabatha Thompson. *Ad Astra* v. 22 pp20–23 Summer 2010.

According to Scott Keepers, structures and mechanical systems manager at The Boeing Company, NASA's prime contractor for the International Space Station, the incredible part of the collaborative spirit is four or five teams working in tandem toward a common goal and getting it to come together finally, Thompson writes. One of the best recent examples of this improvisation came during NASA's October 2007 Shuttle mission to the ISS. Astronauts were on a spacewalk to reposition one of the solar arrays that powers the station when they encountered a snag. Each solar array wing comprises two photovoltaic blankets that arrive folded accordion-style. These blankets mechanically extend and retract along a mast angled to capture the energy of the sun. Guide wires extend the length of the blankets via eyelets similar to those that hold line along a fishing pole. As the blanket unfolded, it caught the sunlight, creating a reflection so strong that neither mission control nor the space station crew could see the operation clearly. The writer explains how the crews on the ground and in orbit collaborated to solve the problem.

NASA and the USA Space Transportation System. Richard V. Simpson. *Antiques & Collecting Magazine* v. 115 pp38–41+ October 2010.

During the three decades of NASA's Space Transportation System (STS), the technology and science of space travel have expanded beyond expectation, Simpson observes. The space shuttle is the most complex machine ever constructed and is assisting with the establishment of the International Space Station, the world's biggest orbiting laboratory. The writer outlines the history of the STS. Collectibles associated with space travel are pictured and described.

NASA Shoots the Moon. Richard Talcott. *Astronomy* v. 37 p23 October 2009.

The writer describes two NASA missions to the moon. The *Lunar Reconnaissance Orbiter* (*LRO*) is to send back more information about the moon than any previous mission. For at least a year, the *LRO* will compile high-resolution 3-D maps of the lunar surface, survey the moon at many wavelengths, and explore regions close to the poles where the sun never rises or sets. The *Lunar Crater Observing and Sensing Satellite* (*LCROSS*) promises to reveal whether water ice exists in permanently shadowed craters close to the moon's polar regions; NASA will smash *LCROSS* into the moon and a shepherding satellite will observe the crash and analyze the debris plume as it flies through.

Earthbound. Hanna Rosin. *Atlantic Monthly* v. 306 pp 27–28 September 2010.

In April, President Barack Obama confirmed that he will adhere to plans to retire the space shuttle program, Rosin notes. The space shuttle itself is an engineering marvel, and it has facilitated many scientific advances, as well as deepening knowledge of the galaxy, but it has generally failed to ignite the national imagination.

Astro vs. Astro: NASA Veterans Debate Commercial Shift. James Asker. *Aviation Week & Space Technology* v. 172 p35 July 19, 2010.

According to Asker, astronaut alumni are discussing the possibility of turning the development and operation of human spacecraft over to the commercial sector. Veteran astronaut James Voss gathered signatures from 24 former colleagues, including Buzz Aldrin, Norm Thagard, and Owen Garriott, on a letter to Senator Barbara Mikulski endorsing the safety of commercial spaceflight. The letter states that, by focusing on a simple spacecraft intended only for low-Earth orbit, commercial groups will avoid the complexity of the space shuttle and the more extreme environments encountered by vehicles designed for further exploration.

Shear Magic. Irene Klotz. *Aviation Week & Space Technology* v. 172 pp22–23 December 13, 2010.

Following two near-perfect Falcon 9 launches and the successful orbital operation, re-entry, and parachute landing of its first Dragon capsule, Space Exploration Technologies (SpaceX) is confident of reaching the International Space Station in 2011, Klotz writes. Encouraged by Dragon's maiden flight, Elon Musk, CEO and chief technical officer for SpaceX, said that he will ask NASA to combine objectives set out for the remaining Commercial Orbital Transportation Service missions and allow a docking at the station during its next fight. If successful, SpaceX would be ready to begin station cargo runs before the end of 2011.

Boost Phase. Irene Klotz. *Aviation Week & Space Technology* v. 172 p40 November 1–8, 2010.

NASA is upping the ante for commercial human space travel with $200 million in the offing for firms to flesh out or flight-test technologies, an effort that will strengthen a new market being pioneered by Bigelow Aerospace to operate leased outposts in orbit. In anticipation of commercial launching services, real estate tycoon Robert Bigelow, owner and president of Bigelow Aerospace, is overseeing a 185,000-square-foot expansion of his Las Vegas–based plant to make inflatable habitats for companies, educational institutes, and countries wishing to stake a claim in space.

Leave It to Us. Guy Norris. *Aviation Week & Space Technology* v. 172 pp22–23 August 9, 2010.

While NASA considers its heavy-lift launch options, Space Exploration Technologies (SpaceX) is unveiling ambitious concepts for a family of Falcon X and XX super-heavy-lift vehicles that it claims could supply the basis for the first commercially based road map to Mars, Norris observes. For the transition from this planet to Mars, however, SpaceX thinks nuclear thermal is the preferred propulsion means for the piloted part of the mission, while solar-electric power could be employed to transport supplies. The firm revealed its exploration vision at the American Institute of Aeronautics and Astronautics Joint Propulsion Conference in Nashville, Tennessee.

NASA's Plan Is Not Sustainable. Jonathan Penn. *Aviation Week & Space Technology* v. 172 p74 December 6, 2010.

With the American economy stuck in a high-unemployment, low-growth pattern, a newly resurgent Republican Party is getting ready to take over the House of Representatives pledging serious deficit reduction, and Americans as a whole display virtually no interest in the current U.S. manned space program, Tang reports. Under these circumstances, the bid to return humans to the Moon or take them to Mars is doomed, given probable NASA budgets. Without radical rethinking and restructuring, the space program will not attain the public support it requires to survive the current economic doldrums and will be unable to help the United States to remain a technology leader in the 21st century.

Map of the Moon. Yuankai Tang. *Beijing Review* v. 53 pp36–37 August 12, 2010.

In 2010, China's first lunar probe Chang'e 1 finished the full map of the moon based on the data it collected, completing its primary objective. Scientists used millions of pieces of data obtained by the laser altimeter to make a full moon digital elevation model. China's moon map has yet to be released and will be put up for sale shortly. It is believed that this is the most comprehensive and accurate lunar map ever published.

The Moon in Close Up. Yuankai Tang. *Beijing Review* v. 53 pp42–43 July 22, 2010.

Scientists working for China's lunar probe program have designed lunar rovers and there are plans to land a rover on the Moon in 2013, the writer observes. Chang'e 2, the lunar orbiter, will prepare for the landing of the lunar rover, including selecting a suitable landing site. The rover's "arms and legs" are highly intelligent and can probe and gauge its environment as the vehicle proceeds. The bottom of the rover carries radar equipment which can "see" several kilometers under the Moon's surface. It is also outfitted with seven sets of apparatus, including an astronomical telescope, thus enabling the first astronomical observations made by a rover on the Moon.

Lost in Space. Paul M. Barrett. *BusinessWeek* pp66–73 November 1–7, 2010.

In February, the Obama administration suddenly canceled an over-budget space program dubbed Constellation that was meant to take Americans back to the moon for the first time since 1972 and then on to Mars, Barrett notes. For 30 years, NASA has flown the Space Shuttle, constructed and maintained the International Space Station, and supervised unmanned scientific probes, but nobody appears sure of where Americans should go next in space. Attempting to protect jobs and existing contracts, Congress has slowed the Obama reform push without wholly stopping it, leaving NASA mired in what James E. Ball, an agency program manager in Florida, refers to as a time of sustained ambiguity.

Space Oddities and Austerities. Michael Brooks. *New Statesman* v. 139 p11 December 20, 2010/January 2, 2011.

The writer considers Britain's space exploration interests in the face of the country's strict austerity measures. He notes how NASA has been squeezing the space shuttle program's resources for a number of years in the face of funding restrictions and envisages British scientists engaged in space exploration having to do the same.

For Mission to Mars, a New Road Map. Kenneth Chang. *The New York Times* pD4 June 8, 2010.

At a workshop held in Galveston, Texas, in May, members of NASA study teams presented their ideas on how the agency can implement the Obama administration's proposed space policy, Chang observes. The administration's policy is based on the premise that the development of new technologies can speed space exploration at lower costs. The plans presented at the workshop place a heavy emphasis on orbiting refueling stations, which would reduce rocket sizes, and a nuclear-powered ion engine, to propel a manned mission to Mars. NASA is also planning a series of robotic missions to the moon, asteroids, and Mars, to gather data for use in planning future manned missions. Some observers, however, expressed fears that the programs could fall victim to budget cuts, which have historically affected NASA technology programs.

Absent a Moon or Mars, Recreating Space 65 Feet Under the Sea. Kenneth Chang. *The New York Times* pD3 May 11, 2010.

Astronauts are training for potential future manned missions at an undersea laboratory near Key Largo, Florida, Chang reports. In May, a group of six astronauts descended to the Aquarius laboratory in the fourteenth mission in the 9-year-old NASA Extreme Environment Operations (Neemo) program. During their two-week stay, the astronauts performed simulated spacewalks, operated a crane, and performed other tasks that might be involved in building a habitat on another planet. They can adjust the buoyancy of their diving suits to simulate the one-sixth gravity of the moon or the three-eighths gravity of Mars, and for half of the mission, a 20-minute time lag in communications will mimic the delay between Earth and Mars. The mission is proceeding despite the current turmoil surrounding NASA's human spaceflight program, which saw the Obama administration scrap the Constellation plan to send astronauts back to the Moon.

Rocket Men. Jeremy McCarter. *Newsweek* v. 154 pp48–51 November 9, 2009.

Barack Obama will need to capture Americans' imaginations if he wants to advance U.S. space exploration, McCarter declares. Given that the most necessary work at the moment is to inspire the American people with the possibilities of space exploration, a special role falls to America's creative artists. When writers and filmmakers do great work about space, McCarter writes, they retrain the public eye heavenward.

From Lunacy to Launchpad. Mark Jannot. *Popular Science* v. 276 p6 January 2010.

In this editorial, the writer discusses the development of the space-tourism industry, stating that the New Space industry is now on the verge of viability. Private companies have 40 orbital flights scheduled between now and 2014, and a blue-ribbon advisory panel last fall was strongly in favor of using private rockets to get astronauts to low-Earth orbit. In 2004, when the journal published a series on the foundations of the industry, the prospect of private space exploration seemed fanciful at best, but today, it appears inevitable.

Space, Inc. Sam Howe Verhovek. *Popular Science* v. 276 pp34–41 January 2010.

The writer describes Virgin Galactic's bullet-nosed rocket, *SpaceShipTwo*, at the Scaled Composites aircraft factory, in Mojave, California. *SpaceShipTwo* is the 60-

foot-long, feather-winged vehicle Virgin was preparing to unveil in December. The craft is part of its grand plan to bring space travel to a slightly broader swath of humanity than has ever been able to contemplate it before. Other plans of Virgin Galactic on space missions are discussed.

Robots on Mars! A. J. S. Rayl. *Reader's Digest* v. 176 pp162–69 June/July 2010.

The writer discusses the missions that the Spirit and Opportunity Mars Exploration Rovers have been doing since landing on Mars in 2004.

Jump-Starting the Orbital Economy. David H. Freedman. *Scientific American* v. 303 pp88–93 December 2010.

According to Freedman, NASA's plan to abandon the manned spaceflight business may jumpstart the orbital economy. Under the plan, NASA will provide seed money to start-ups and then buy tickets to the space station on their rockets. This plan is in many ways an attempt to return NASA to its 1960s glory days by making it a true research and development agency again. It is still too early to tell whether the private sector can deliver a safe, reliable orbital vehicle, but there have been positive signs. The plan's biggest potential payoff would be the opportunity to drive costs of a flight to orbit down low enough to create an orbital economy.

Next in Space. Mark Strauss. *Smithsonian* v. 41 pp88–89 July/August 2010.

Strauss presents a guide to future probes and observatories due to be launched into space by NASA, the European Space Agency, and the Japanese Aerospace Exploration Agency.

Launch Signals New Space Race. Andy Pasztor. *Wall Street Journal* ppB1–B2 December 9, 2010.

In a crucial test for Space Exploration Technologies Corp. (SpaceX), the first commercial space capsule, Dragon, was successfully launched and recovered, Pasztor reports. Poised to carry cargo and perhaps eventually U.S. astronauts to the International Space Station, the Dragon capsule reached roughly 17,000 miles per hour before surviving a fierce re-entry and landing gently in the Pacific.

One Small Step for Science? Matt Mahoney. *Technology Review* (Cambridge, Mass.: 1998) v. 112 p120 September/October 2009.

The celebration of the 40th anniversary of the Apollo 11 mission revives the debate over the scientific merit of manned space exploration, Mahoney contends. There has not been a lunar landing since 1972, and as the glories of the Apollo mission have been remembered, there has been a call for renewed commitment to manned space exploration. Extracts are presented from a 40-year-old dispatch by the legendary journalist Victor Cohn detailing a contentious and surprisingly public fight between scientists and NASA officials in what should have been the agency's finest hour, which appeared in the *Technology Review* issue immediately following the successful lunar landing.

Q&A: Buzz Aldrin. Brittany Sauser. *Technology Review* (Cambridge, Mass.: 1998) v. 113 pp26–27 July/August 2010.

In April, President Obama flew to Kennedy Space Center in Cape Canaveral, Florida, to unveil details of his new strategy for NASA and the future of U.S. spaceflight, Sauser observes. Sitting next to Obama on Air Force One was Buzz Aldrin, who in July 1969 became the second man to walk on the Moon. The seating arrangement was appropriate, given that both men share a common goal for the nation's space program: reaching Mars by the mid-2030s. In this interview with Sauser, Aldrin discusses his ideas for the future of U.S. human spaceflight.

No One Can Hear You Scream. Ivan Hansen. *The Walrus* v. 7 pp19–20 September 2010.

As human technology progresses and astronauts travel farther from Earth, human vulnerability has become the space program's greatest restriction, Hansen writes. Canada has made its most significant nonhuman contribution to space medicine with Devon Island, a massive, uninhabited island in Canada's High Arctic that is home to the Haughton Mars Project, the only polar research station whose purpose is to support space exploration.

Invaders of Mars. Katharine Gammon. *Wired* v. 18 pp54–55 June 2010.

For the past five decades, humans have been sending orbiters and landers to Mars, Gammon observes. Pieces of manmade hardware are scattered all over the Red Planet, some the fragmented results of less-than-soft landings, others observant little robots still transmitting their observations back to Earth. A map of where man-made robots have been on Mars is presented.

Index

About the Editor

A lifelong space buff, **Christopher Mari** was born and raised in Brooklyn, New York. He has worked as a writer and editor since completing his formal education at Fordham University. His nonfiction and short stories have been published in *The Absent Willow Review*, *Citizen Culture Magazine*, *Current Biography*, and *Midnight Times*, among other periodicals. In addition to working on this current volume, he has served as editor or coeditor of four additional books in The Reference Shelf series: *Space Exploration*, *Global Epidemics*, *The American Presidency*, and *The Next Space Age*. He lives with his family in Queens, New York.